Economic Theory and the Construction Industry

Economic Theory and the Construction Industry

Patricia M. Hillebrandt

Third Edition

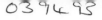

palgrave

First edition 1974
Reprinted four times
Second edition 1985
Third edition 2000

Published by
MACMILLAN PRESS LTD
Houndmills, Basingstoke, Hampshire RG21 6XS
and London
Companies and representatives
throughout the world

ISBN 0–333–77478–7 hardcover
ISBN 0–333–77479–5 paperback

A catalogue record for this book is available
from the British Library.

This book is printed on paper suitable for recycling and
made from fully managed and sustained forest sources.

Transferred to digital printing 2002

Printed and bound in Great Britain by
Antony Rowe Ltd, Eastbourne

To Mary Greaves, who made me write it

Contents

Preface to the First Edition

The need for an analysis such as is presented in this book became clear to me while I was working as an economist in Richard Costain Ltd and later in the National Economic Development Office. My experience there convinced me that there were special problems of the construction industry in which application of the basic concepts of economics would be very helpful, but that this industry was in certain respects so different from others that the theory needed tailoring before meaningful application was profitable.

I had no opportunity to develop any theoretical approach to the construction industry until I was faced with the problem of introducing economics to students in the then Bartlett School of Architecture, now part of the School of Environmental Studies at University College, London.

More specifically, this book grew out of a one-year course given to MSc students taking as one of their options the subject of building economics. These students came from many backgrounds: building, management, civil, mechanical or structural engineering, architecture, quantity surveying and economics. Most of them had considerable experience and some of them were themselves teachers in polytechnics or other institutions of higher education. They were taking the course in order to bring a new dimension into their thinking about the construction industry, and my problem was therefore to present economic theory in a way which would seem relevant to them and yet with sufficient depth for the students to be able to adopt its inherent logic as part of their own mental equipment which would be of value no matter with which aspect of construction they were later concerned. As the course developed, it became increasingly clear that there were many aspects of the industry which did not fit into classical economic theory, at least not without considerable adaptation. Slowly the theory that was taught began to be more specific to the construction industry. Hence the book which emerges is one which aims to present economic theory to those concerned with construction and at the same time to interest economists in the special problems of the construction industry.

The book draws on the parts of economic theory which are useful and relevant for this purpose, concentrating on those areas where there is some analysis particularly appropriate to construction. In order that it shall be readable by non-economists without necessary reference to other texts, a general analysis of a situation is sometimes required. As a result, economists will find it possible to skip over Chapter 11 on market equilibrium and types of market situation. Non-economists who wish to delve deeper into the subject would be well advised to broaden their understanding by reading any good economic textbook, particularly for analysis of factors of production, general equilibrium and macroeconomics, areas which this book hardly touches.

As far as I know, this is the first book to treat the construction industry from this point of view. It is, however, a field in which many questions remain unanswered. Some of them can be considered from theoretical standpoints alone, but most require applied research in the field to test the validity of hypotheses. It is hoped that this first tentative outline of the map will tempt researchers to study selected areas in greater detail.

Without the stimulus of discussion in seminars with successive groups of MSc students, particularly those in the years 1971–2 and 1972–3, this book would not have been written. Many of my colleagues have given help on specific chapters. I should like to mention particularly Professor G.L.S. Shackle, who read part of the chapter on the price determination for a single project and gave help and encouragement for my use of his degree of potential surprise function. I am greatly indebted to the three friends and colleagues Professor Marian Bowley, Professor Duccio Turin and Mrs Margaret Bloom, who read and commented on the whole draft. As a result of their valuable criticisms and suggestions many amendments have been made and some parts completely redrafted. The responsibility for the deficiencies and errors which remain is mine alone.

PATRICIA M. HILLEBRANDT

Preface to the Second Edition

The climate in which the construction industry in the United Kingdom and other Western developed countries functions in 1984 is very different from that in 1973 when this book was first written. The early 1970s were a period of boom in the UK and to some extent across Western Europe whereas in the early 1980s the level of output of construction is at considerably lower levels and policies of government have altered. Problems generally have changed from having insufficient resources to having unemployed resources in the developed world.

Some of the earlier parts of the book which set the scene for the theoretical sections have been rewritten, notably much of Chapter 2 and parts of Chapter 3 and the whole of Chapter 7. In updating, the scene has been broadened to include comment and examples from developing countries, the Western European countries and the USA as well as the United Kingdom thus reducing the somewhat parochial flavour of the earlier edition where concern with the UK construction scene was paramount. The theory is, after all, applicable across international boundaries.

Few alterations have been made to the theory chapters, apart from one or two corrections and clarifications. However, some of the references and the bibliography have been updated.

PATRICIA M. HILLEBRANDT

Preface to the Third Edition

Since the last edition of this book in 1985, there have been many and substantial changes in the construction industry. For the theory set out in this book, the most important are an increased internationalisation of the industry, change in the services demanded from the industry and in their method of procurement and, in some countries, the increased power of the client.

As a result, chapters on the way the industry functions, including the product demanded and the process by which demand is met, have been substantially rewritten. A new chapter has been included on the demand for refurbishment, repair and maintenance to rectify an inexplicable omission in earlier editions. In addition, two new chapters have been included to show how theory can help on issues touching on the broader choices facing the construction firm and the far-reaching choices, linked to the construction industry, in the management of the economy. Most of the theory remains as before, although greater emphasis has been given to aspects of theory which have come to be more generally used in the last few years and some of the lessons from theory as to the nature of the industry have been revised in relation to the changes which have taken place.

I should like to thank friends and colleagues who have helped me in writing this revision and, in particular, Jim Meikle, Ted O'Neil and George Ofori who have kindly read and commented on it as a whole. The remaining errors are mine alone. I should also like to thank Anna Wilson and my son, John Hopf, for so patiently guiding me through the mysteries of the word processor in this, my first major, belated attempt at using it.

PATRICIA M. HILLEBRANDT

Abbreviations and Acronyms

AC	average cost
BOOT	build, operate, own and transfer
BOT	build, operate and transfer (sometimes as BOOT)
CFR	Construction Forecasting and Research Ltd
CIB	International Council for Research and Innovation in Building and Construction
CIC	Construction Industry Council (UK)
CIOB	Chartered Institute of Building
D&B	design and build
DETR	Department of Environment, Transport and the Regions
DLC	Davis Langdon Consultancy
DL&E	Davis Langdon and Everest
DLSI	Davis Langdon & Seah International
EU	European Union
FSU	Former Soviet Union
GDP	gross domestic product
GNP	gross national product
HMSO	Her Majesty's Stationery Office (UK)
ICOR	incremental capital output ratio
IMF	International Monetary Fund
ISIC	International Standard Industrial Classification
JRF	Joseph Rowntree Foundation
MC	marginal cost
NEDC	National Economic Development Office (UK)
NIC	newly industrialised country
NJCC	National Joint Consultative Council (UK)
ONS	Office of National Statistics (UK)
PFI	private finance initiative
RICS	Royal Institute of Chartered Surveyors
TC	total cost
TG29	Task Group 29 (of the CIB)
UK	United Kingdom
UN	United Nations
USA	United States of America

Part One
Introduction

1
The Nature of Construction Economics

Scope of subject

Construction economics consists of the application of the techniques and expertise of economics to the study of the construction firm, the construction process and the construction industry. To understand this definition and the scope of the subject, it is necessary to know what economics is about, what constitutes construction and why the construction industry deserves a special branch of economics to itself.

A practical definition of economics is that 'Economics is the science which studies human behaviour as a relationship between ends and scarce means which have alternative uses.' (Robbins, 1935). This definition has not been surpassed in so far as the economics of the firm and the industry are concerned, though others abound. On this definition, economics is essentially about the allocation of scarce resources. If resources are not scarce there is no problem. If a resource has only one use there is no problem. If there is only one objective there is no problem. This definition does not indicate the coverage of, for example, national income accounting and related model building for a whole economy or the work of Lord Keynes (Keynes, 1936), describing the interrelationships of the total economic system. Keynes himself described economics as 'a method rather than a doctrine, an apparatus of the mind, a technique of thinking which helps its possessor to draw correct conclusions' (Keynes, 1921). This opens up the way for an economics of many other topics, such as family life and crime (Groenewegen, 1987). It also opens the way for an economics of construction.

The construction industry is an industry whose product is the

3

services necessary to produce durable buildings and works. In the International Standard Industrial Classification (ISIC) (UN, 1968) construction covers general and special trade contractors primarily engaged in contract construction. Also included are units of enterprises primarily engaged in construction work for a parent enterprise. The activity of construction covers new construction, alterations, repair and demolition of buildings and civil engineering works. It also includes assembly and installation on site of prefabricated integral parts of buildings or works, including departments of manufacturers engaged in this activity. However, it is convenient for present purposes to include not only the contractors as specified in the ISIC but also the professions which design and cost the product and manage its production. A definition agreed internationally at a conference on developing countries is:

> The construction industry comprises all those organisations and persons concerned with the process by which building and civil engineering works (following the activities identified in the International Standard Industrial Classification (ISIC) (UN, 1968)) are procured, produced, altered, repaired, maintained and demolished. This includes companies, firms and individuals working as consultants, main and sub-contractors, material producers, equipment suppliers and builders' merchants. The industry has a close relationship with clients and financiers. (CIB TG 29, 1998)

In spite of the diversity of the construction industry, in the nature of its product, in the types of organisations and in the process by which production is organised, it is nevertheless one industry. For certain purposes, it is convenient to regard it as a number of sub-industries and it certainly embraces a great range of different markets. There is a school of thought that regards the concept of one construction industry as counter-productive and would prefer to consider the project as the basis for analysis, broadening out to embrace the totality of contributions to the creation of the project including 'external linkages – and potential innovators from beyond "construction" altogether' (Groak, 1992, 1994). There are instances when it is essential to concentrate on the nature of the project to understand the process of production. There are also circumstances in which it is convenient to place the spotlight on one segment of the industry. However, construction has characteristics which, separately, are shared by other industries, but in combination appear in construction alone, justifying its industry status. These characteristics fall into four main groups: the physical nature of the product; the structure of the industry, together with the organisation of the construction process; the determinants of demand; and the method of price determination.

The final product of the industry is large, heavy and expensive. It is required over a wide geographical area and is, for the most part, made especially to the requirements of the individual customer. Most of the components of the industry are manufactured elsewhere by other industries. It is largely these product characteristics which determine the structure of the industry, including the large number of dispersed contracting firms and the usual separation of design from construction which has such important repercussions. The nature of the product, together with the structure of the industry it encourages, also means that each contract often represents a large proportion of the work of a contractor at any one time, causing substantial discontinuities in the flow of work. The work of the contracting part of the industry involves the assembly of a large variety of materials and components.

Demand on the construction industry is, on the one hand, for investment goods and, on the other hand, for their maintenance, repair, rehabilitation and eventual demolition. The investment goods are of several types:

- as a means to further production, for example, factory building;
- as an addition to, or improvement of, the infrastructure of the economy, for example, roads;
- as social investment, for example, hospitals;
- as an investment good for direct enjoyment, for example, housing.

The determinants of demand for these categories of product are different and need separate analysis. This is the subject of Part Two of this book. All demand is affected, though in different degree, by the ups and downs of the economic cycle and by actions of government, either directly as a client of the industry, or by the way in which it runs the economy.

Because of the physical nature of the project, the structure of the industry and the characteristics of demand, the method of price determination is usually a discrete process for each project and for each piece of work subcontracted (see Chapter 14). This is normally by a process of tendering, by some form of negotiation or by a mixture of the two. These processes may be applied whatever the product or service provided, for example, with design and build, the construction and management of the building and finance for the building. However, because of the diversity of the services provided by the industry (see Chapter 8) price determination often involves complex procedures.

There is some work, where the developer and the contractor are the same organisation and hence there is no overt price determination for the project. The price which the developer charges for the finished product, whether it is for sale or rent, is influenced by many factors other than the construction cost, such as the price of land, the price of capital and the system of taxation. These projects are therefore not covered in this book, except in so far as it is necessary to do so to understand the determinants of demand on the industry.

The specific characteristics of construction discussed above justify its designation as an industry and also the need for its special study. It remains to consider the nature of construction economics.

Construction economics may be regarded as a branch of general economics. Ofori (1994) has carefully considered whether this is justified. He comes to the conclusion that '[c]onstruction cannot be regarded as a *bona fide* academic discipline. It lacks a clear indication of its main concerns and contents and a coherent theory. It is not recognised as a distinct part of general economics.' He calls for a conscious effort to develop an appropriate body of knowledge to justify the description as a branch of general economics. This book is an attempt to make a contribution to the development of a coherent theory of construction economics.

The contribution of the economist so far to the problems of the construction industry has been slight. It has been mainly in the field of macroeconomics of the construction industry and its place in the economy and to a lesser extent the economics of the firm. There is very little economic theory of the construction project or the construction site. There have been and are some exceptions to this situation. Marion Bowley's publications of the 1960s are classic studies of the industry (Bowley, 1960, 1966). Others, either from the discipline of economics or a subject area related to construction, have produced substantial publications which have made significant contributions to the thinking on economic aspects of construction, for example, Ball (1988), Ofori (1990, 1993), Raftery (1991), Groak (1992) and, for the future, volumes by Ive and Gruneberg (Ive and Gruneberg, 2000 and Gruneberg and Ive, 2000) will do much to improve the position.

The situation has been confused by the designation of quantity surveyors as 'building economists'. When they claimed this name in the 1960s, they were not economists as defined as persons who practise the science of economics as discussed above. They were essentially concerned with prices of buildings and of the inputs which are involved in their construction. Since then the training of quantity surveyors has changed and degree courses which prepare them for their profession generally contain some economic theory. This has encouraged some of them to broaden their horizons to economic aspects of the construction industry. One can start as an economist and learn about construction or vice versa.

I believe there is a subject of construction economics, albeit in its infancy, and that it is recognised as such by the construction community. It will not be recognised by economists until more of the main stream economists become interested in the construction industry and understand its peculiar characteristics. At the moment there is little interest.

Relationship to other subjects

Just as economics overlaps with other subjects, so too does construction economics. Whatever decisions are taken in the design of buildings or works, in their construction or in the management of the firms which are involved, they are nearly always taken after consideration of factors other than the purely economic. Indeed, in any decisions taken in a construction firm, economic factors and management implications will nearly always be considered together. Similarly, the macroeconomic decisions which affect the industry must take cognisance of non-economic matters.

Hillebrandt and Cannon sought to discover what various disciplines, management theory, sociology and finance for example, had to offer to the management of construction firms (Hillebrandt and Cannon (eds), 1989). In a linked study (Hillebrandt and Cannon, 1990) they explored how large UK contractors take their business decisions, and investigated the awareness of managers in contracting firms of ideas from various disciplines. Although there were exceptions, most managers in construction firms were unaware of the potential contribution of economics, management science, sociology and other disciplines to the improvement of management. The survey demonstrated, however, that there are a number of areas where greater awareness would be beneficial. These were categorised in three groups:

- The first group answers the question 'What should be done and how?' It includes the formulation of company objectives and strategies and method of corporate planning. This aspect of management theory is of great importance and has strong links with economic theory.
- The second group answers the question 'Have all the relevant factors and their effects been taken into consideration and assessed?' It provides a logical framework of the relevant factors in the choice between various options. It includes the theory of diversification, of international policy, and of finance, all of which are part of, or closely liked to, economic theory. It also includes some theory of the management of organisations.
- The third group deals with aspects of corporate and human activity and implicitly invites a comparison with practice. It includes more sociological approaches to the structure of organisations and labour management. Much of this theory is relevant but mainly as a background to other ideas of more direct and immediate relevance.

Amongst all the disciplines the strongest link is between economic theory and organisation theory. Some of the fundamentals of modern theory were developed by economists interested in management. Simon questioned profit maximisation as the objective of entrepreneurs and developed the theory of bounded rationality (Simon, 1959). Coase and later Williamson with their concern with explaining the need for organisations, led to the concept of transaction costs (Coase, 1937; Williamson, 1975). Baumol too was concerning himself with organisations, and, with others, developed the theory of contestable markets (Baumol *et al.*, 1982). Porter has been influential in developing an understanding of competitive strategies for businesses and nations (Porter, 1990). This prolific area of study, linked with economic theory is sometimes described as managerial economics. Social science research on the construction industry makes some use of conventional organisation theory but mainstream economics plays very little part in the concerns of researchers in construction (Betts and Lansley, 1993). While some management concepts are referred to in this book, the emphasis is on more traditional economic theory of the firm.

2
Some Basic Concepts in Economics

There are some concepts used by economists which it is appropriate should be explained at the outset, in order that the flow of the argument in later chapters is not interrupted by a digression to explain their meaning. First, the meaning of market as opposed to industry is considered. There follows a discussion of various types of cost, and of price and profit. The nature of marginal analysis and the concepts of supply and demand are then explained and, finally, two ways of handling economic data for analysis are described.

For further definitions of terms, the reader is referred to the index where page references to definitions and explanations of the use of a term are shown in heavy type.

Construction markets

The nature of the construction industry has already been discussed in Chapter 1. When considering many of the matters of interest to economists, however, it is not the industry but the market which is relevant. A market, in the economist's sense, is any organisation whereby the buyers and sellers of a particular commodity keep in touch with each other and determine the price of the commodity. The organisation required may be of any degree of formality. It may, at one extreme, be of the type of the modern stock exchange where transactions are undertaken by computer and the buyers and sellers do not have to meet in any geographical location or, at the other extreme, like a cattle market where buyers and sellers meet with their product to determine, on the spot, the price for each animal. In contracting it means the whole mechanism of the selection of the contractor and the fixing of the price at which he will provide his services. The sellers in any particular market in construction are the group of firms whose services to provide various products are more or less substitutes for each other in terms of the type of expertise required. The products produced in a particular market may be similar and be substitutes for each other, for example housing, but, if the skills required are similar, they may have quite different uses, for example, small office blocks and schools. Geographical location may also define a market. Whereas there is probably one market for large, sophisticated office blocks in the UK, for smaller, simple offices there are several markets, depending on accessibility to the smaller contractors who undertake that type of work. Rarely in the UK would a contractor travel across the Pennines for a small job. The sellers in any market compete against each other for work. The nature and size of these construction industry markets and the factors which determine their limits need to be considered in some detail.

Type of finished product is one area which limits a contractor's involvement. There is a major difference between undertaking some large civil engineering works and say, low rise housing. There is less difference between, say, offices and factories. Recent analysis found that differences in contractor competitivenes is greater for different contract sizes than for different contract types (Drew and Skitmore, 1997).

Size of contract is clearly a major determinant of the number of firms who can undertake work. A large contract requires more of all inputs than a small contract and only some of the total contractors in the country have these inputs available to them. Capital and management are particularly important in this connection.

Complexity of the product is another determinant of the potential competitors. A complicated building can be constructed only by firms having control over the technical expertise required, either in-house or by buying in expertise. If expertise is brought in from outside, it is still necessary to manage the contribution of the outside experts and that is better done from a position of knowledge. It should theoretically be possible to measure the degree of complexity but the practical difficulties are great. Probably the nearest readily available approximation is cost per square metre for building of a given type. If size and complexity are combined, the number of firms in that market would be lower.

The effect of location on the number of firms in a market is an important factor in determining the size of market. One reason why firms do not go outside a certain area of operation is that the costs of transport of materials, plant and men, as well as that of managers' time travelling, become excessive in relation to other costs. The larger and more expensive the project, the less the proportion of transport costs in the total, so that it becomes feasible to venture farther afield for larger projects than for smaller projects. For very large projects the firm will set up a local office, so that the transport costs decrease but the overheads increase.

Another determinant of the market has recently become very important, namely, type of service which is being supplied and the related contractual arrangement. In the UK, at least, contractors have for many years specialised in say, design and build or management contracting. Now, however, there is great diversity in the service provided, including provision of a building or works and its management and maintenance over a period of many years. The development of arrangements for private finance for construction products formerly entirely in the public sector has effectively selected another variety of market type.

Apart from what markets it is theoretically possible for a firm to be in, the firm itself may decide, as a matter of policy, to limit the breadth of the market in which it wishes to operate. At the same time, clients may decide that a firm cannot satisfactorily work in a particular market. A firm which has grown in size may be deemed unsuitable for its previous small work because of the difficulty of getting the personal attention of the now, very busy, managing director.

Markets in the construction industry should therefore be defined in terms of the total demand for a particular identifiable service which is not a close substitute for other services outside this market. Relevant parameters include degree of complexity and size, geographical area and type of contractual arrangement. The total number of firms interested in work of this defined type are referred to as being 'in a particular market'.

Firms may, indeed do, operate in more than one market and very large firms will undertake work across most of the spectrum of type of work, complexity and geographical location, and to some extent, size of work. A large firm will sometimes do small work, usually through a separate part of the business or from regional offices. Fluctuations in demand for a particular type of work are greater than in the demand on the industry as a whole. Therefore, contractors who have a broad range of work reduce fluctuations in demand levels. Another reason is that the skills of the main contractor on building work are basically those of managing a complex assembly process, so that the end use of the project is not critical.

Construction firms

The word 'firm' in economics means any entrepreneurial unit. It can be a single person, a partnership, a small company, a public limited company or a gigantic multinational organisation.

Types of cost

It follows from the nature of economics, as a subject dealing with scarce resources and their use to maximise a benefit of some kind, that the use of a resources for one purpose implies its withdrawal from another purpose. The cost of using a resource in use A is the lost opportunity of its employment in B. Hence the real cost of using the resource in A is the lost opportunity or 'opportunity cost'.

As an example, consider the opportunity cost of capital. If money capital is invested in a painting business, the opportunity cost of its use there is the return which could be obtained on that capital with the same risk in another industry – say plant hire or even watchmaking. Similarly, if all the resources used to build a school could have been used (in the same quantities and proportions) to build a health clinic, then the opportunity cost of the school is the health clinic. In practice, the resource mix is unlikely to be the same for two buildings and therefore some other way has to be found of expressing the lost opportunity. In a perfectly functioning competitive economic system, the price paid for all the resources is equal to the opportunity cost, since the price of each commodity is equal to its value in all its various uses. This is so because in each use successive units of a good are employed up to the point at which the value of the output attributable to the use of the last unit of the good equals its price. In such a system, therefore, opportunity cost equals the price paid for resources.

In practice the economic system is not perfectly competitive and does not function without some frictions, and hence opportunity cost and price paid for resources may diverge. When there is less than perfect competition a resource may not always be used up to the point at which the value of the output attributable to the use of the last unit equals price, and this is one source of error in the use of price as a measure of real cost. Probably even more important is that prices do not respond immediately to the pressures of supply and demand. This 'stickiness' of prices means that, for example, the price of skilled manpower, i.e. the wage rate, may not equate demand and supply so that at the ruling price there is some short-fall in supply and the process of allocation of supply has to be done by some other method in addition to price. This may be rationing, first come first served, or perhaps by some non-money or at least non-wage inducements to labour to satisfy certain types of demand. This 'stickiness' in wages is due to a number of factors. Changes in wages for a skill tend to be related to wages in other industries and changes in the cost of living, and these factors exert a restraining influence on the speedy adjustment of wage rates to scarcity. Trade union influence, too, largely because it involves negotiating through elaborate institutional machinery, tends at times to be a delaying factor in adjustment.

Two other major cost concepts are long-term costs and short-term costs. Long and short term are not fixed periods of time but vary according to the matter under consideration. In general, the short term is a period so short that there are certain factors which cannot be altered. Thus the costs of producing additional output of sheet glass from an old plant with much maintenance requirement and substantial labour and material cost will be high. The cost of producing additional output of sheet glass from a new, technically efficient plant may be much lower. But the new plant would take, say, five years to build. In the short run (five years in this case) the costs are high, in the long run low.

A further pair of costs are social costs and private costs. Social costs are costs to the community; private costs are costs to the individual or group of individuals. The nature of these and other costs will be illustrated in some detail by an exploration of one of the best examples of various types of cost, namely the decisions on the desirability of using industrialised building for housing in the 1960s. At that time there was an urgent need to build more housing and the level of construction output was increasing in nearly all types of work. There was also a shortage of certain inputs, notably skilled manpower, but also of some materials. A report on the state of the industry and the likely future demands indicated that there was uncertainty as to whether the industry would be able to meet the demands upon it. It was feared that construction prices might rise with harmful effects on the economy as a whole. If the industry failed to meet demand the economy would suffer (NEDC, 1964). It was against this background that industrialised construction methods were encouraged.

In the early 1960s the price of a dwelling unit built by industrialised methods, which use more capital and less labour than traditional methods, was higher than the price of a unit built by traditional techniques. The exception was high rise construction over five storeys where the industrialised was cheaper. At this time the amount of industrialised construction was relatively small, although a boom was just beginning. Observers of the industry in this period were of the opinion that the relative prices were not representative of the real costs or opportunity costs because of the stickiness of wage rates in a time of shortage of skilled manpower.

Moreover, it was expected that, in the long run, costs of industrialised construction would fall as, with an increasing production, the high initial capital costs were spread over a larger output. It was anticipated that there would also be a reduction in costs due to increased experience with the new technology. Thus short-term costs were thought to be higher than long-term costs. It was also expected that in the long run wage rates would rise, thus giving industrialised building, with its lower labour usage, an even greater long-term cost advantage.

It was perceived, therefore, that the money costs of industrialised building were higher than the opportunity costs and that the short-term costs were higher than the long-term costs. Given the environment of the time referred to above, it is not surprising that the conclusion was reached that the private costs of industrialised construction were higher than the social costs. These costs take into account the use of resources, the long-run view and the broader concerns of the effect of actions on the whole economy, because society has a continuing responsibility, not only for the construction industry, but also for the economy as a whole.

This example, of decisions taken in the UK some 35 years ago has some relevance to the future in view of the renewed interest in industrialised construction. It also bears a strong resemblance to the sort of analysis which developing countries should undertake when considering appropriate technologies. This subject is dealt with fully in Chapter 16.

The concept of social cost has been broadened in the last few decades to include the idea of sustainable development. This may be defined, in relation to the construction industry, as creation of buildings and works in the present without detriment, or with minimum detriment, to the ability to produce similarly in the future. This implies *inter alia*, minimal use of finite resources; prevention of waste of resources; and use of the process of construction, including the organisations and institutions, in such a way as to ensure their continuing health and ability to produce in the future. Thus, social costs must include the costs of misuse of resources of all types. Governments are becoming increasingly aware of the importance of ensuring sustainability and proper environmental protection. Action currently tends to be piecemeal but in the future is likely to be much more interventionist.

Energy saving is one area where government has acted in relation to building. An indication of the change in attitudes is given by the introduction in 1994 by the US Commerce Department of *Environmental Accounts*, starting with subsoil assets like oil, gas and coal (US Department of Commerce, 1994). As oil is a finite asset, it is desirable to keep its use as low as reasonably possible. However, the individual consumer has no great incentive to conserve energy, unless it is cheaper to do so, because he is not interested in the oil situation beyond his lifetime. But society is! Because society has a long-term interest, some governments give subsidies for energy-saving measures such as roof insulation.

Because the economist is concerned with the value of a resource in an alternative use, if the resource no longer has an alternative use, that is, it has no opportunity cost, it ought not to be considered in the decision-making process. If, for example, research has been devoted to developing a particular piece of machinery and it appears that the likelihood of success is slim, in considering whether to continue with the research, the labour which has been spent in the past is irrelevant because it now has no opportunity cost. All that is relevant is the cost of the resources (including opportunity cost) to be spent in the future compared with the benefits likely to be derived from them. Costs incurred in the past, which cannot now be altered are known as 'sunk costs'. They may be capital costs or variable costs, for example, a highly specific building which has no resale value or a quantity of manpower. Not all costs incurred in the past are sunk costs. Past expenditure on a factory may be partly offset by the sale of the factory or its use for other purposes.

Other types of cost, for example, transaction costs, fixed costs, variable costs and so on are defined in Chapter 10.

Price

Price is the rate at which exchange may or does take place and it applies to all resources and factors of production; thus the price of cement is so many pounds a tonne, the price of labour is wages, the price of capital is interest. Price is usually expressed in money terms, but in a barter economy, where there is no money, there is also a price, for example grain in terms of knives. Price has an important part to play in the workings of the economic system as the mechanism which balances demand and supply.

It is sometimes easy to confuse price and cost. That is because what is a price to one person may be a cost to someone else. A cement producer will sell cement to a contractor at a figure which is a price to him. It is a price to the contractor too, in the sense that it represents the exchange terms dictated by the market. The contractor, however, will view what he had to pay for the cement as a cost, because it is the cost of one of the inputs to his construction project.

Profit

Profit is the revenue obtained by a firm in excess of its costs. Since costs in the economic sense include a normal return on capital and on entrepreneurial ability, that is, a return sufficient to keep the capital and entrepreneur in the industry, firms can stay in business making no profit in the economic sense, although not in the accounting sense.

Marginal analysis

Marginal analysis is very important for economics and its use will be explained in detail in later chapters. It deals with small changes in the total as a result of some other change. In the case of marginal cost, for example, economists ask the question: 'What difference will it make to total cost if production is increased?' The difference in the total is the marginal cost.

The concern with the marginal differences goes right through economic analysis, and marginal cost is just one of many marginal concepts used by the economist. He also asks the questions: 'What will be the difference in my total satisfaction if I have another £1000 of income per annum? What will be the increase in the revenue from sales if I increase my sales by, say, 20 units?' He asks similar questions about changes to that total as a result of falls in sales, decreases in production, etc. Thus, for any projected change he is able to assess the decrease in total benefit on the one hand,

and the increase in total benefit on the other hand, and hence whether the final benefit is larger or smaller than it would have been without the change. If he assesses this marginal change for very small changes in the variable, he can find his optimum position.

The presentation of marginal analysis, as it is called, may be done graphically, as in this book, or by differential calculus, which gives more accurate results and can be used for a greater number of variables. The latter does, however, require a familiarity with differential calculus and a mathematical approach and way of thinking not assumed in this book.

Demand and supply

A detailed analysis of the nature of demand and supply forms the major part of this book. The two concepts are, however, so closely interlinked that it is necessary to refer to supply curves in the demand section and hence some brief explanation is appropriate here.

The amount of any commodity demanded at various price levels can be expressed graphically as in Figure 2.1 below, where the demand curve for a commodity is line *AB*, such that the lower the price, the greater the amount demanded. *CD* is the supply curve for the same commodity, showing that the lower the price the less will be the amount on offer. The curves cut each other at *E*. At price *OF* the amount demanded is *OG* and the amount which is on offer or supply is also *OG*. *E* is the point of equilibrium and the point at which exchange will take place. If the price were greater than *OF*, more would be on offer than demanded and suppliers would tend to lower their price until the amount demanded equalled the amount they wanted to sell; while if the price were less than *OF*, the amount demanded would be greater than the amount on offer and the would-be purchasers would be willing to pay more up to a price of *OF* at which the demands at that price were met by the supply at that price.

Methods of simplification

In order to study the workings of the complex and vast economic system, economists have developed methods of simplification to enable them to analyse the data. Here only two are mentioned. The first is used to study the effect of change in one variable. In real life everything is changing all the time and by observation it

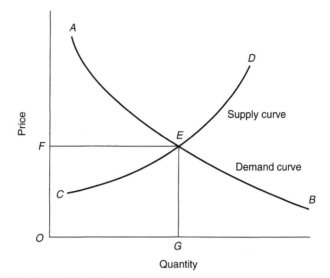

Figure 2.1   Demand and supply curves

is difficult to see what effects arise from any particular cause. The economist tries to work out the effect of one change by assuming that nothing else is causing change at the same time. He assumes 'Other things being equal' or *ceteris paribus*. At a later stage in this investigation he can relax these simplifying assumptions one by one and see what the different effect would be. Thus we can, for example, study the effects of a training levy on construction firms, assuming in the first place that the levels of demand and of resource availability remain as before. Both of these assumptions may be false in reality but both assumptions are necessary to bring the problem down to a manageable size.

The other main method of simplification is aggregation, that is, lumping together a great mass of individual items and looking at them as a whole. This is what we are doing in speaking of the amount of investment or gross national product (GNP). This is what the Minister of Finance does when he tries to see how the economy is progressing and what he should do about the level of taxation. Great use was made of this method by Keynes (1936). By working with great blocks of income, consumption, investment and so on, he was able sufficiently to simplify the system to be able to study the interrelationships of all the main parts of the system at once.

3
The Construction Industry and the Economy

Importance of size

The construction industry is important partly because its output is large and therefore that it is a significant part of the economy. The gross output of the construction industry is the value of all the buildings and works produced by the industry in a given period of time, normally a year. In the world as a whole it is probably about 10 per cent of Gross National Product (GNP), that is 10 per cent of all the goods and services produced, or of the order of US$3,000 billion in 1997. There is a considerable difference between various types of economy and geographical locations. Davis Langdon Consultancy (DLC, 1997) estimated that in 1990 the percentage shares were: Western Europe 30 per cent; Asia 28 per cent; North America 25 per cent; Eastern Europe 7 per cent; South America 5 per cent and Africa, Middle East and Oceania just under 2 per cent each. These estimates are similar in general conclusions to those by Drewer (1999) who estimated that developed market economies accounted for 78 per cent of global construction output in 1990. In Europe, gross construction output was around 10 per cent of GDP in 1997, marginally less for the EU and slightly more for countries of the former Soviet Union (DL&E, 2000). Some countries in Asia have a very high level of construction output, for example China where construction output is well over 20 per cent of GNP. By contrast, in the USA construction accounts for only about 9 per cent of GNP and in the UK too it is low. (DL&E, 2000; DLSI, 1997)

The net output of construction is the same as value added, that is, it is the value of the contribution of the construction industry itself to the production of building and works and excludes all the inputs of other industries, such as building materials and plant and equipment. The relationship between gross output and net output depends on the technologies used and the relative prices of inputs. If, for example, technology is fairly simple and the cost of labour is low, net output is likely to be low compared to gross output. This is the case in China. The value of net output is normally about half gross output in developed countries and less than half in developing countries. Low (1991) calculated that over the fifteen year period from 1970 to 1984, North America had the highest regional output followed by Western Europe and East Asia. 'The massive market of the USA in North America, the congregation of advanced industrialised countries in Western Europe and the significant presence of Japan in East Asia appear to be plausible reasons that account for their respective positions as the top three in the league table.' (Low, 1991, p. 66). Whatever measure is used, an industry which produces such a large component of GNP is of great significance for the economy.

Investment industry

An even more important reason for the construction industry's importance in the economy is that it produces investment goods. This means that its products are wanted, not for their own sake, but on account of the goods and services which they can create or help to create. Investment is vital for the wellbeing of any economy and, although it may be postponed for a while, if it is not undertaken, the economy will deteriorate. In most countries construction provides about half the gross domestic fixed capital formation, that is, half the production in the economy which is invested. The remainder of investment is plant and machinery, for example in factories, and vehicles and other transport goods. In some countries certain information technology products are also regarded as investment.

It is clear that, say, factory building is an investment because it is used to create other commodities. In a different way it is also true of school building where the building is not required for its own sake, but as a place in which to produce education. Housing can be regarded as the place where accommodation is produced. This stretches the argument rather far, however, and, although housing has other attributes of an investment good, it may be regarded as directly consumed.

Most investment goods have a long life, often considerably longer than the life of the person or organisation which creates them and they are often sold to another user when the first owner or user no longer needs them. Another characteristic of investment goods is that they are usually very expensive so that their cost is high in relation to the income of the purchaser. For the individual consumer, for example, the purchase of a house will usually entail expenditure of several times his annual income. Similarly, the erection costs of a factory by a manufacturing firm will be a large expenditure in relation to the running costs of production and in relation to the annual income derived from it. Consequently the products of the construction industry, with the exception of repair and maintenance, are paid for out of capital, that is the purchasing power that has accumulated in the past but not been used in the past. This capital may belong to the owner of the construction product or, more usually, be borrowed from elsewhere.

By reason of the long life of construction products, the stock of products is large in relation to the annual production. Small fluctuations in the demand for the stock of buildings and works will have very large repercussions on the demand for buildings and works created by the industry.

Links with government

In nearly all countries, government, central or local, or acting through some public or semi-public organisation, has an important influence on the construction industry. This occurs in a number of different ways of which one of the most important is the effect of government actions to control the whole economy, for example, the rate of interest, the amount of public sector expenditure and, related to this, the system of taxation. Many of these matters will be discussed in more detail later in this chapter.

A second factor to be considered is the way that government arranges for the construction of infrastructure and other goods regarded as public goods, such as roads, water supply or schools. At one time, in most countries such provision was managed by government providing the product directly and acting as client, and sometimes as contractor, for the required structures. The reasons for the direct involvement of government in the provision of such products may be divided into three main groups. In the first place, there is the problem that many people will want to use the asset (although the potential users are not necessarily an identifiable group) but that the product is too expensive for any one group of persons to pay for. Roads fall into this category. In the case of roads, there is the further problem that they take up space on land occupied by a number of different persons, at least some of whom are unlikely to be willing to sell. Some form of compulsion is necessary so that government must be involved in any case. A second reason is that the provision of the goods or services provided by the asset is regarded by government as its responsibility; state provision is then the obvious solution. This category includes schools, hospitals, a clean water supply and electricity. In some countries these services are privately provided but, in one way or another, are usually regulated by government.

A third important factor is that the production of goods or services arising from the assets, for example, power, is most economically undertaken on a large scale with high capital investment but very low running costs, so that the cheapest provision could be provided by just one firm. Thus there is a natural monopoly situation and one of the ways of controlling a natural monopoly is by the direct ownership, or at least control, of government. Another way is to break up the monopoly and allow competition.

In the last twenty years or so there has been a move away from government ownership towards greater participation of the private sector. This change is closely related to a political swing towards a strong belief in the merits of the market economy. This change of policy has occurred across many developed countries and some developing ones. The governments of South East Asia, especially Malaysia, have been at the forefront of privatisation of infrastructure (Ofori and Rashid, 1996; Tam and Leung, 1999). Many operations previously regarded as firmly in the public domain, such as power, water supply and railways have been privatised. Arrangements have been developed for methods of finance of infrastructure and public buildings such as hospitals, which draw on private finance as well as, or instead of, public finance. This change has been dramatic in many countries, including the UK, but nowhere as dramatic as in the former Soviet Union where, encouraged by the IMF and the World Bank, the whole structure of government ownership and control has been dismantled and replaced by privatised firms. Unfortunately the economies of these countries are in a serious state, perhaps partly because it was too sudden a change which included the privatisation of several firms which could never be viable in a market economy.

At the same time, the shift towards private enterprise has been helped by technological developments. These have, for example, enabled the monitoring of the usage of public assets, for example roads and water, and made possible the provision of certain services with much less capital investment than previously, for example telephones. The latter has enabled the introduction of competition into services which had previously been natural monopolies.

Thus, although government is the client for substantial amounts of construction work, it now operates much more as an enabler with such devices as the Private Finance Initiative (PFI) in the UK which provides for the use of private finance, sometimes in combination with some public money. This does not mean that government has no influence on the way the industry operates. On the contrary, in some ways it is exercising more control because of the breadth of scope of the new arrangements covering design, construction, and operation of the finished asset. In many cases the asset is transferred to government at the end of a certain period of years so that it needs to ensure that the project is viable in the long term.

It is not only as a client or enabler that government controls the industry but also as a regulator of the construction process. Through policy on planning it controls the supply of land for development, through building regulations it enforces certain minimum building standards and it legislates on employment conditions and contracts.

Construction and management of the economy

In the last ten or twenty years detailed management of the economy by government, which was normal practice in the 1950s and 1960s, has been abandoned, to be replaced by a firm determination to control inflation with as high a level of activity as can be sustained, largely using the rate of interest and fiscal policies. Many of the controls available to government, such as those on bank lending, have been discontinued. In an increasing number of countries, control of the rate of interest is now the role of the Central Bank, rather than the government, which means that there is a danger of uncoordinated policies and makes any return to 'fine tuning' of the economy more difficult. The major reason for the change is that governments were often unsuccessful in achieving their objective of an economy working near its capacity without wide fluctuations in output. This disappointment with past achievements coincided with the ideological change towards belief in relatively unfettered market forces producing the best outcomes.

There are, however, several weapons which governments have to control the economy. Government controls the level of its own spending and, using taxation and borrowing powers, it can determine whether the way it finances its spending is likely to increase or decrease the level of demand and the likely effect on output. If the economy is working at or near its overall capacity, government may decide to decrease its own spending or, if that is not feasible, increase taxation to restrict the level of demand in the private sector. Whichever course of action is adopted, it will have considerable effects on the demand for construction, especially as it is usually easier to postpone capital expenditure than to cut current spending. If, on the other hand, the economy is in recession, government can take the reverse action to boost the economy and this will increase the demand on the construction industry.

The rate of interest is one of the weapons regularly used by government to control the economy and is currently extensively used by governments or Central Banks as a means of control, with special emphasis on regulating the rate of inflation. Changes in the rate of interest can be used in ways similar to government spending, borrowing and taxing as described above.

The construction industry is substantially affected by changes in the rate of interest, both directly and indirectly. Firstly, an increase in the rate of interest directly affects the industry by increasing the cost of borrowing for contractors, so that those who have a substantial overdraft suffer badly and may be forced out of business. The importance of this factor varies from country to country. Most contractors in developing countries are very dependent on banks for their working capital, whereas, in the UK, because of the traditional project financing arrangements, contractors are often able to undertake projects with low or negative working capital.

Secondly, the interest paid by clients on money they have borrowed for the project increases so that the total cost to clients of construction projects rises. It is not unusual that in a period of rising rates capital investment projects are cancelled or postponed. At the same time, with rising mortgage rates, householders will postpone investment in housing. Overall the workload of the industry will decrease.

Lastly, a rise in interest rates has a more indirect effect on the industry, in that it reduces spending power over the whole economy.

There will be a fall in the demand for manufactured goods and hence the need to build new factories. Consumers' confidence and purchasing power will fall leading to an unwillingness to spend, not only on new housing but also on maintenance and improvement of housing.

This rather doleful picture can of course be reversed by a reduction in interest rates. It is apparent that fluctuations in construction output follow inevitably from changes in interest rates.

Government may encourage certain types of development in order to benefit the economy. In many South East Asian economies, especially those of Hong Kong and Singapore, the development and real estate business provides revenue for government, and its activities generate further prosperity. Government is therefore supportive of its role.

Another example of government intervention is the introduction of incentive schemes and development of infrastructure in order to attract direct foreign investment for the benefit of the economy. This is known as 'nation marketing'.

One of the features of the construction industry is that it is more labour-intensive than many other industries, especially manufacturing. This makes it an attractive industry to use to influence the level of demand and output, and hence employment, in the economy as a whole. This is a possibility because when employment in construction is increased, the workers spend their wages on products of other industries thus increasing the demand for other products and thus, providing there are no bottlenecks, output and employment. This employment, in its turn, generates more employment elsewhere and an upward spiral is started. If there is overload on the economy it is possible to use a similar process to dampen down the activity in the economy. Ways in which this process may be used to help developing countries and examples of its use in other circumstances are discussed in detail in Chapter 16.

Whether or not the industry is deliberately used to affect the level of activity in the economy, because of its size and its high level of employment, its state of health is likely to have an effect on the whole economy. Similarly, it would be expected that construction would be affected by the economy. In other words, they interact with each other.

Fluctuations in construction output

Fluctuations in construction output, which, in their more extreme form, are called construction, or more usually, building, cycles are endemic in the industry. In part they are caused by fluctuations in the economy as a whole and in part by the nature of the construction product. There is a long history of interest by economists in building cycles, not because they were interested in the building industry but because they wanted to understand and explain the trade cycle generally. A US study undertaken for the Committee for Economic Development in 1952 covers the subject well, looking not only at the causes of instability but also at possible solutions (Colean and Newcomb, 1952).

Parry Lewis (1965), in a painstaking analysis of building cycles in Britain from 1700, has found that there are distinct long-term fluctuations in construction activity, often regional in character, and finds the 'key ... in population, credit and shocks...'. The shocks he has in mind are, for example, bad harvests or war, which will have repercussions on the natural increase in population, on migration, etc. They will also affect the monetary sector, often through the balance of payments. Population and credit as exogenous variables can be taken into account in making long-term forecasts of construction activity, for example of housing and education, but shocks cannot by their very nature be taken into consideration in the forecasting process. Yet clearly their effect on construction and the effect of construction on the economy through the level of total production and through the demand for money can be as important as the effect of the economy on the construction industry.

Kafandaris (1980) has examined various early work on the subject finding references to building even before the analysis of cycles which developed from the middle of the nineteenth century. He is very critical of the failure to try to understand the industry and to distinguish between, for example, house-building and the construction of infrastructure in describing or discussing reasons for construction cycles.

For the UK, Ive and Gruneberg (2000), in comparing the cycles between 1950 and 1973 with those since 1973, found that contraction and expansion amplitudes in the construction sector have increased and that the downs have been greater than the ups indicating a rather stagnant industry. In addition construction cycles have shown greater amplitude than the cycles in GDP. Fluctuations in construction output would occur, to some extent, even if no factors external to the industry were influencing the situation. The long life of some construction products and the large stock of buildings and works of different ages would themselves generate fluctuations in the need for construction, although external factors have a very strong influence on when need can be converted into economic demand.

The perception of the magnitude of fluctuations in the construction industry by the persons and organisations working in it is often greater than the overall statistics of output would seem to justify. There are several reasons for this. First, most of the firms in the industry are small and work in a relatively small geographical area. Because projects are produced where they are required and because they are relatively large and spaced out, any one contractor may experience a substantial fall in the demand for work in his catchment area. If he specialises in a particular type of work the fluctuations in work available to him will be even greater. There is a further matter. The contractor will be aware in his marketing of new orders rather than the output of work as the contracts are worked. New orders are considerably more erratic than output. That partly explains the vociferous complaints of the industry in the UK in the 1960s about stop – go. In fact, that period was relatively stable, as found by Ive and Gruneberg. Total output stopped increasing and then rose again. It did not fall; but the industry felt as though it did. There seems no prospect of eliminating the fluctuations in construction output, although better planning of construction programmes in relation to the capacity of the industry, especially in developing countries, could help to reduce them and efforts should be made in this direction (Hillebrandt, 1999a). Meanwhile, the industry would do well to assume that it has to exist with fluctuations in demand and firms should take whatever defensive action is available to them.

Balance of payments and construction

Historically, construction has been a domestic industry with the firms, the labour and the materials being indigenous. This situation has changed dramatically in newly industrialised countries (NICs) and to a large extent in developing countries also, where there is a demand for modern sophisticated buildings for which the country does not have the materials and not always the skills and management required. Moreover, because these countries cannot afford the buildings and works which they need, many projects are built with international aid. Where the import of materials and other inputs are paid for from outside the country, there is no damage to the balance of payments. In circumstances in which imports have to be paid for by internal funds, the impact on the balance of payments can be serious.

In developed countries, and to some extent in developing countries, the increasing international activities of contractors affect the balance of payments. Increasingly, apparently domestic contracting firms have a foreign shareholding and for large projects joint ventures are very usual. No study is known of the effect on the flows between countries of these activities.

Part Two

The Demands on the Construction Industry

Introduction

Demand is regarded by economists as the requirement for goods or services which the customer is both able and willing to pay for. It must not be confused with need which, for buildings and works, may be defined as the difference between an accepted standard of provision and the extent to which the stock reaches that standard.

The determination of demands on the construction industry is complex. This is partly due to the characteristics of the products of the industry: notably, their size, cost and long life; the fact that many of the products are investment goods; and the complexity of the process. There are five major requirements that must be favourable before demand can be created:

- that there is a user or potential user for the building or works in the short run and/or in the long run;
- that some person or organisation is prepared to own the building or works;
- that a person or organisation is prepared to provide finance for the construction of the product and for its ultimate ownership;
- that some person or organisation is prepared to initiate the process;
- that the environment and external conditions in which the above operate are favourable.

There are thus four persons or institutions involved: the user, the owner, the financier and the initiator. To these one could add government as having a substantial influence on the environment in which demand creation takes place. In practice, two or more of these roles can be combined though the combinations vary for different product types and over time (Hillebrandt, 1984)

The many categories of construction demand in terms of product characteristics include the following:

- New projects to meet a previously unsatisfied need, for example, a new airport or the alleviation of overcrowding in housing. The demand for such projects may arise at any time and in many countries the need, as opposed to the economic demand, is almost endless.
- New investment to extend the provision of certain types of buildings or works, for example, an increase in factory building to meet increased demand for output or of water purification plant to meet the needs of increased population. The accelerator principle is important in this connection. (see Chapter 5)
- Replacement of existing stock of buildings or works, although often in a different form, Examples are housing, industrial buildings and schools. These projects will be undertaken when the useful life of the existing stock is over for any reason.
- Alterations to a building to change the use to which it is put, for example, conversion of warehouses to housing.
- Rehabilitation and improvement of the existing stock which may be necessary at irregular intervals but shorter than the life of the building.
- Repair and maintenance to keep buildings and works in good order. This is needed at regular intervals or continually.

It will be seen that the frequency of the occurrence of demand for these items varies greatly from one to the other. Moreover, most of the demands are postponeable and depend on the state of the economy and other factors. Inevitably this causes fluctuations in levels of demand.

In the UK and in some other countries, the firm line between publicly funded and privately funded projects is disappearing. This is partly due to attempts to measure usage of social type projects so that fair recompense can be made for the use of private finance as in some Private Fnance Initiatives (PFI), described in Chapter 8. It is also due to the establishment of hybrid organisations which are part private and part public. Changes in categories are particularly important for housing and social type construction.

In the chapters which follow, the factors affecting demand for various types of product are considered. The first is the construction of housing in Chapter 4. This chapter begins with the demand curve of the individual for units of housing. It is possible to go back further to the way in which the individual decides how much housing he wants at various prices. This involves an understanding of indifference curve analysis and is dealt with, for those interested, in Appendix A. Indifference curves are a useful tool, not only for housing, but also for other analysis. Chapter 5 deals with the type of goods desired as a means to an end of producing some other good or benefit, illustrated by industrial and commercial building. Chapter 6 covers the demand for social type goods, generally built for the collective enjoyment or benefit of a large group of persons, for example, schools and reservoirs. Then, the demand for rehabilitation and repair and maintenance is considered in Chapter 7. Finally, in this part of the book on demand, the way in which demand is put to the industry is described in Chapter 8.

4
Demand for Housing

The construction of housing enables a benefit to be available for direct consumption. In a minority of cases, the client of the industry is the user, as in privately commissioned housing. Most housing, however, is either produced by a private developer in advance of orders by users or by a public, quasi-public or private organisation for renting (or sale) to users. The private developer is known as a speculative builder – speculative in the sense that there is uncertainty as to whether the dwelling will be sold. In this sense at least a large part of the output of manufacturing industry is speculative. The manufacturer of toothbrushes does not know his clients when he produces the brushes. However, the sheer size of the purchase of a house and the fact that it is a postponable capital transaction renders speculative housing a more uncertain business than the manufacture of toothbrushes.

Public authorities and quasi-public organisations normally construct dwellings only when there is a clear demand at the price at which it is prepared to let them. This price may be lower than the market price and contain an element of subsidy, whereas the private builder will usually sell or rent at the market price and will produce dwellings only when he anticipates that he will make a profit. In spite of the differences between the various types of client, all are dependent on the consumers assessment of the desirability of housing compared with all the other goods and services they could buy with their available resources and hence on how much they are prepared to buy at various prices. In Appendix A the underlying basis of consumer demand is analysed with the help of indifference curves. In this chapter it is assumed that the individual consumer has decided how much housing he is prepared to purchase at various prices.

Individual's demand curve for housing

In Figure 4.1 on the horizontal axis is measured the quantity of housing in notional units per annum. On the vertical axis is the price of these notional housing units. Then, the demand curve for an individual for housing units will take the approximate form of, say, D_1. The curve slopes downwards to the right and is convex to the origin, since the more units of housing the individual has, the less he values further units of housing. Thus, with a price of 20, the individual will demand 7 units of housing, as shown by point A, at a price of 15 he will demand 10 units (point B), and at a price 10 he will demand 15 units (point C).

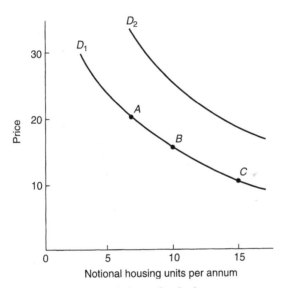

Figure 4.1 The demand curve of the individual

Concept of notional units of housing

The measurement of housing in terms of notional units of housing is a conceptual tool. Housing is not homogeneous. Any system by which one tries to add up the standards of light, space, plumbing, heating, electricity, aesthetics – to mention only a few – must be based on assessments of their relative values which are matters of opinion. In addition, apart from the characteristics of the house itself, there are external factors: neighbourhood amenities, open spaces, level of industrial land use and so on (see Cheshire and Sheppard, 1998). The most practical method of arriving at the relative value of various amenities is to express them all in money values, but this is of little assistance for present purposes, as it is often the very money values (or prices) which are under study. Economists who have worked on housing and tried to assess elasticities of demand (see below) have been forced to consider what unit they are able to use. One of the most useful for the United Kingdom is the rateable value of property, and this was used by Clark and Jones (1971) in their study of demand for housing.

Price of housing

The price of housing in Figure 4.1 has been expressed in some arbitrary unit of money. It represents the amount out of his annual income which the individual is prepared to give up for housing. It may take the form of rent to a public authority, a quasi-public organisation or a private owner, of interest payable on borrowed capital to purchase a property or, if he has used his own capital, of the interest which he has forgone on his investments in order to transfer his capital to property ownership. This is the real rate of interest or the opportunity cost of the use of his capital in this way.

In the case of public authority housing, the relevant price is that charged to the individual after deduction of subsidies. Subsidiary costs of housing related directly to the amount consumed should be included: for example, repair, maintenance and service charges on private-sector flats sold on a long lease, local taxes, repair and maintenance of the dwelling. For houses purchased by the individual, there are complications arising from the taxation position and from the effects of inflation and price changes in the value of assets. These are dealt with in some detail below.

Effect of taxation

In those countries where there is tax relief on house purchase, the effects can be substantial. It is the rate of interest which is actually paid after all tax allowances which should be considered or, in the case of purchase out of capital, the interest after all tax deductions which is the forgone interest.

Effect of costs of buying and moving house

The actual price paid for the house is not the only capital outlay. The transaction costs of owning include estate agent's fees, the costs of finding the right dwelling and the cost of moving. In addition there are the costs of rehabilitation and of necessary repairs and furnishing. Hendershott and Shilling (1982) regard the transaction costs as proportional to house value and as being amortised over the length of stay. For simplification here they could simply be added to the cost of the house.

Effect of changes in the value of capital assets

If the value of money is changing or if there is a change in the price of dwellings or other relevant assets, some adjustment ought to be made to the rate of interest considered as the price of housing.

Table 4.1 shows the factors which have to be considered for purchase of a house both out of borrowed capital and out of own capital. It is assumed in the case of borrowed capital that the loan is not available for use except on house purchase, or in other terms that the increase in the value of the capital in an alternative use (d) is nil. Considering the borrowed capital situation first, the net rate of interest is the actual rate payable minus tax relief, i.e. $c=a-b=5$ per cent in the example given. If the value of the dwelling increases, this has to be offset against the net rate of interest and the effective rate becomes $f=c-e=-5$ per cent.

Table 4.1 Example of calculation of effective rate of interest as the price of housing

	Borrowed capital	%		Own capital	%
a	Rate of interest payable on loan	8	g	Rate of interest in alternative use	8
b	Tax relief, say,	3	h	Tax payable	3
c	Rate of interest net of tax, i.e. $a - b$	5	i	Rate of interest net of tax, i.e. $g - h$	5
d	Increase in value of capital in alternative use	0	j	Increase in value of capital in alternative use	5
			k	Effective foregone rate of interest, i.e. $i + j$	10
e	Increase in value of dwelling	10	l	Increase in value of dwelling	10
f	Effective rate of interest, i.e. $c - e$	−5	m	Effective rate of interest, i.e. $k - l$	0

If, on the other hand, the purchase is made out of the person's own capital, then the rate of interest in its alternative use has to be considered minus the tax payable, i.e. $i=g-h=5$ per cent. However, this capital was increasing its value too, so that the effective rate of interest which is foregone is $k=i+j=10$ per cent. Then, the effective rate of interest for house purchase is the effective forgone rate minus the rate of increase in the value of the dwelling, i.e. $m=k-l=0$ per cent.

If this rate f for borrowed capital is used to calculate the price of housing in relation to other commodities, the price becomes negative, i.e. the individual is paid to consume housing units, and demand would be infinite; but this clearly does not conform to reality. The reasons are as follows:

(i) It is difficult to obtain credit to purchase a house unless income is fairly high, and then it is easy only for the first house.

(ii) There are other forms of capital investment besides housing which include some hedge against inflation, e.g. equities.

(iii) There are costs of consumption of housing other than rent, notably property taxes, maintenance and cleaning (see above).

(iv) House purchase is regarded as a long-term matter and except in periods of rapid price increases, few persons wish to move at frequent intervals. Although the effective rate of interest at the moment of purchase may be negative, it is a matter of uncertainty whether or not it will remain so, as there may be a change in interest rate, taxation rates or inflation.

(v) Most people do not calculate the cost of housing in terms of real costs but in terms of actual current outgoings. In the case of the adjustment of the rate of interest for taxation, the time period is relatively short. However, in the case of adjustment for inflation, the time period is extremely long – the period until the house is sold. Even when it is sold the family must live somewhere and will not normally wish to reduce its standard of housing. A similar house will cost a similar price and again the capital gain is not realised (unless the mortgage is increased). In other words, the payment to buy housing comes at the end of the period of consumption of housing, or with a reduction in the consumption of housing due, for example, to children leaving home.

For these reasons, it is not very satisfactory to deal with the effects of price changes in the capital assets by making an adjustment to the interest rate payable. It is probably more appropriate to confine the price to the net rate of interest after tax and to consider the effect of price changes as a shift in the demand curve.

Shift in the demand curve

The demand curve shows what amount of a good would be demanded at various prices. It applies to a particular point in time so that all variables other than price are constant. If there are changes in the circumstances of the individual over time, the demand curve will change in position or shape or both. This is known as a shift in the demand curve. A change in circumstances may be caused by any of an infinite number of causes, such as a change in income, a change in his taste for housing by reason of advertising for example, a change in his ownership of capital, etc. It is suggested too that there would be a shift in the demand curve of an individual if he came to anticipate a change in the value of housing for purchase (although only for those individuals considering house purchase).

A change in income is particularly important for housing demand because housing expenditure accounts for such a high proportion of total income. As income rises there would usually be a shift of the demand curve upwards to, say, D_2 in Figure 4.1. The effects of income changes are analysed in depth in Appendix A.

Capital ownership will clearly have a considerable effect on the demand for housing, since it increases the resources available for the consumption of housing. A discussion of its treatment is also included in Appendix A. Clearly, however, any change in the ownership of capital by an individual would tend to shift his demand curve.

Total demand curve

So far the analysis has been confined to the nature of the individual's demand curve for housing. It is now possible to see how, using the individual demand curves, a curve showing the total demand for housing units can be derived. Suppose in Figure 4.2 that (a), (b) and (c) represent the individual demand curves for housing of three persons. These three curves can be summed in (d) to see how much all three consumers together would demand at given prices. Thus at price 30 the demand would be 4.5 units from (a) plus 2 units from (b) plus 7.5 units from (c) to give a total of 14 units in (d).

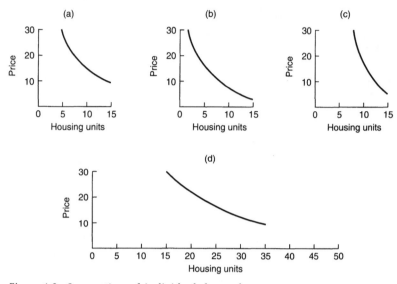

Figure 4.2   Summation of individual demand curves

Just as three can be summed so, in theory, can the demand curves of the total number of consumers and the total demand curve obtained, although of course changing the scale on a horizontal axis. The total demand curve will slope down to the right. It may well be convex to the origin since most individual demand curves are of this shape. However, in the total of individual curves there may be some which are not convex and these could be sufficient to make the total demand curve other than convex – perhaps approximating to a straight line.

It is clear in the derivation of individual demand curves from indifference curves and price (see Appendix A) that they represent what the individual will actually require at various prices. Thus total demand curves, sometimes called economic or effective demand curves, show what the community is able and willing to pay for and should not be confused with need, which is discussed below.

Price elasticity of demand

In order to describe the demand curve more precisely, the concept of elasticity is useful. Elasticity of demand expresses the percentage change in the amount demanded in response to a percentage change

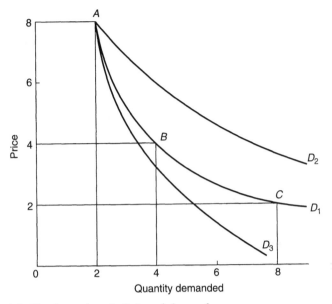

Figure 4.3 Varying price elasticies of demand

in price. It is perhaps most easily understood in terms of the total amount spent on a commodity. If, with a fall in price, the amount demanded increases more than proportionately to the price change, so that the total amount the consumer spends on a commodity rises, then the demand is said to be elastic, that is, it stretches (hence elastic) much in response to price. If, however, with a fall in price the amount demanded increases less than proportionately to the price fall, so that the total amount the consumer spends on a commodity falls, then the demand is said to be inelastic. The neutral position arises where the amount the consumer spends is the same whatever the price. In this situation demand is neither elastic nor inelastic and, as will be seen below, when the measurement of elasticity is discussed, is expressed as elasticity 1 or $e = 1$. Demand is likely to be elastic if there are close substitutes for the product (e.g. flats as opposed to houses) and inelastic if the commodity has no close substitutes (e.g. all housing).

The demand curve of $e = 1$ is the shape shown in Figure 4.3 as D_1. No matter what the price, the area of any rectangle drawn under the curve representing the amount spent, that is, price multiplied by quantity demanded, is always constant. Thus in Figure 4.3, at

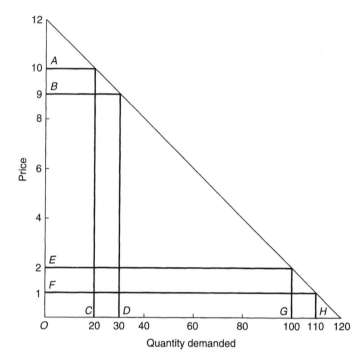

Figure 4.4   Demand curve with slope = −1

A, *B* and *C* and at all intermediate points the consumer will spend 16 money units on the product. This type of curve is described as a rectangular hyperbola.

When the curve is flatter than the curve *e* = 1, then *e* is greater than 1 and the curve is elastic. When the curve is less flat than *e* = 1, then *e* is less than 1, i.e. the curve is inelastic. In Figure 4.3 the curve D_2 has *e* greater than 1 and the curve *D*, has *e* smaller than 1.

More precise numerical values may be put to elasticity. Price elasticity of demand for good *x*

$$= -\ \frac{\text{Percentage change in quantity of } x \text{ demanded}}{\text{Percentage change in the price of } x}$$

The minus sign is simply a convenience of definition to make elasticity positive. Let us consider this definition in terms of a straight line with slope = −1, as in Figure 4.4. Consider a fall in price from *OA* to *OB* or from 10 to 9; demand increases from *OC* to *OD* or from 20 to 30, that is, with a fall in price of 1 from 10 by 10 per cent, the quantity demanded increases by 50 per cent.

Therefore
$$e = -\frac{\dfrac{50}{100}}{\dfrac{-10}{100}} = -\frac{50}{-10} = 5.$$

At a different part of the curve, consider a fall in price from OE to OF or from 2 to 1. Then demand increases from OG to OH or 100 to 110. In this case

$$e = -\frac{\dfrac{10}{100}}{\dfrac{-50}{100}} = -\frac{10}{-50} = \frac{1}{5}.$$

It is clear from this illustration that a straight line, that is, a curve of constant slope, has different elasticities over its length. This is because it is the change in relation to the total which is expressed by elasticity, not the change by itself.

Income elasticity of demand

A parallel concept to price elasticity is income elasticity. Income elasticity expresses the percentage change in the amount demanded in response to a percentage change in income.

Thus, for example, if income increases by 20 per cent and the amount of a commodity purchased increases by 10 per cent, then income elasticity of demand is $\frac{1}{2}$; if the amount purchased increases by 20 per cent, it is 1, and if the amount increases by 40 per cent, then the income elasticity of demand is 2. In broad terms, if income elasticity of demand is less than 1 the commodity is a necessity, and if it is greater than 1 it is a luxury. As the proportion of income spent on housing is high, income elasticity of demand is an important concept for the consideration of the demand for housing. Several attempts have been made to measure it using either time-series analysis, in which the available data on housing demand are analysed in relation to income changes over a period of years, or cross-section analysis, in which housing demand is examined in relation to families living in various income groups in a given short period of time. The latter method has the advantage that the time period is too short for other conditions to have changed substantially.

Assessments of elasticities of demand

The early work on income elasticities of demand for housing was done in the USA, and Reid (1962) drew on the work of Friedman (1957) and others for her book *Housing and Income*, where she makes an exhaustive study of American data distinguishing between 'normal' income and actual income. Normal income is the stable income which potential customers of housing 'have in mind when making decisions'.

She found high elasticities of demand for housing. Muth (1960) also found that both price and income elasticity were high, perhaps as high as 1.0, although he also observed that the lag in adjustment is considerable. Later household level studies (summarized in Quigley, 1979, and Mayo, 1981) provide much lower demand elasticity estimates (often +0.5 or lower) (Goodman, 1989).

In the early 1970s there was a burst of activity in the UK in estimating income elasticities and price elasticities of demand for housing, probably on account of the change in the relationship of the number of households and the number of dwellings which makes some understanding of elasticities essential for forecasting.

Most of the work has concentrated on private-sector housing for owner occupation and refers to expenditure on housing as a whole, which would of course include the cost of land. Permanent income elasticity of demand based on cross-section analysis ranges from 0.6 to 1.0 (Holmans, 1971; Clark and Jones, 1971; Johnston *et al.*, 1972; Vipond and Walker, 1972), largely depending on the consideration given to age of head of household, social class, etc. Time-series analysis tends to show higher elasticities (Whitehead, 1974), probably due to changes in the underlying conditions in the market for housing and in the population structure.

Holmans, Whitehead, and Clark and Jones also worked on price elasticity and found values ranging from 0.26 to 0.6. The price elasticity demand for the dwelling (excluding land) is likely to be considerably less elastic than for housing as a whole by reason of the high proportion of the cost of land.

Recent work in this area in the UK includes that by Meen (1994) who, using time-series data, indicates an income elasticity of 0.8 and price elasticity of 0.4. Ermisch *et al.* (1996) used Joseph Rowntree

Foundation (JRF) Housing Finance Survey data in six UK conurbations in 1988–9 and examined 9,500 cases. They found a price elasticity of owner occupiers of 0.4 and their method gave a less confident income elasticity of 0.5. Both these studies aggregate housing characteristics, whereas Cheshire and Sheppard's model (1998) assesses the effect of neighbourhood characteristics as well as housing itself. They study elasticities of many factors, including, for example, number of bedrooms and WCs; type of property, such as flat, terrace, semi-detached; garage; location of schools; type of street and open amenity land. They find elasticities larger than those found by Ermisch *et al.*, as might be expected for disaggregated attributes.

For the public and quasi-public sector the elasticity of demand for housing by the individuals who are to occupy the dwellings has had relatively little relevance. However, it is important for the building industry that it should have some idea of price elasticity of demand for social housing. It seems that, overall, since the assessment of their demand is made on need and since, as in private housing, the price of land is a large component of cost, the price elasticity of demand for the dwelling is likely to be rather inelastic. It is complicated in the UK by the system of housing benefit by which tenants do not pay the whole rent but the landlord receives it from the benefits office.

It should be noted that in a situation where the maximum price of a new dwelling is fixed by statute or regulation, if price rises, the amount demanded in terms of standards or units of housing (not dwelling units) falls, so that the price elasticity of this housing becomes 1.

Demand for new housing

The analysis so far has been concerned with the demand for housing, irrespective of whether the housing is stock or newly constructed. The demand for new housing will depend on the demand for all housing, the stock of housing and the amount of replacement; while the amount actually built will depend on this demand and the building industry's supply curve.

In Figure 4.5A are the demand and supply curves for all housing, and in Figure 4.5B the demand curve for new housing and the supply curve for new housing both expressed as demand and supply per annum. All curves in these diagrams have been drawn for convenience and ease of exposition as straight lines.

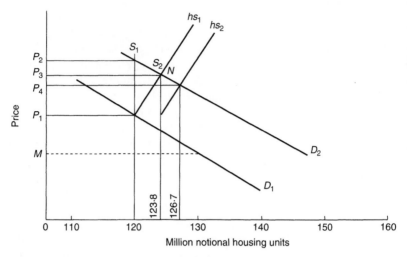

Figure 4.5A   Demand and supply of all housing

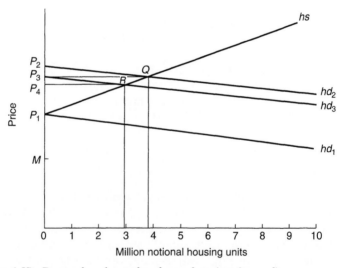

Figure 4.5B   Demand and supply of new housing (annual)

Suppose that at a certain date, say 1 January in year 1, the demand for all housing is shown in Figure 4.5A by the curve D_1 and the stock by S_1 at 120 million notional housing units. Price is P_1, which balances supply and demand.

Assume that the housing supply industry is in a state of equilibrium and at price P_1 is prepared to build nothing for stock, as is shown in Figure 4.5B by its supply curve for new housing *hs*. In such an equilibrium position it might still at that price build for replacement, in which case the supply curve *hs* would shift to the right, but this is not shown in the diagram and in this analysis for the sake of simplification replacement is ignored.

The supply curve of the housing industry is conceived as including the supply curves of private housing developers, local and other public authorities acting as developers, whether letting or selling at a subsidised price or not. Since it is housing accommodation which is being supplied, the price must include an allowance for the cost of land as well as for the actual costs of construction, all expressed as an annual rate in a manner similar to that used for the analysis of demand. In view, therefore, of the high costs of land in many areas, the housing supply curve will be determined to a relatively small extent by the actual costs of construction.

The demand curve for new housing at 1 January in year 1 may be derived from D_1 in Figure 4.5A. From D_1 the demand for new housing at price P_1 is nil since the existing stock meets demand, while at price M it is 130 million units minus the stock of 120 million or 10 million units. This is shown in the annual demand curve for new housing hd_1 in Figure 4.5B.

Assume that for some exterior reason, such as immigration or a rise in incomes, the demand D_1 shifts to D_2 on 1 January of year 2. As at a certain day the stock cannot be increased, the stock remains the same at S_1 and the new price becomes P_2.

Just as the annual demand curve for new housing hd_1 in Figure 4.5B was derived from D_1 in Figure 4.5A, so the housing industry's supply curve for all housing can be derived from the annual supply curve *hs* in Figure 4.5B, adding it on to the stock. Thus the supply curve for all housing during year 2 will be hs_1 in Figure 4.5A. The new point of equilibrium at the end of year 2 will be at N with a stock of 123.8 million notional housing units and a price P_3. For year 3 the analysis is repeated with the supply curve becoming hs_2, the new stock 126.7 and the new price P_4. This process continues until a new long-term equilibrium is established. Whether this is at a price above or below the price P_1 depends on the shape of the curves, for example whether there are any factors pushing up the long-run supply curve. It is clear that any shift in the demand curve will take some years to work its way through the system because the net annual additions to stock are very small in relation to the total stock of housing.

Figure 4.5B shows the same situation viewed from the housing supply industry position. It was already seen that with the initial demand D_1 in Figure 4.5A in equilibrium, there was no building for stock and the price was P_1. When the demand curve for all housing shifts to D_2 in Figure 4.5A there is a new derived annual demand curve for new housing of hd_2 in Figure 4.5B. This cuts the annual supply curve hs at Q with price P_3 and output of the industry 3.8 million units.

In year 3 the industry is faced with another demand curve hd_3, for at price P_3 the demand for addition to stock is now nil (at N in Figure 4.5A, D_2 and S_2 cut) and the annual demand curve becomes hd_3 in Figure 4.5B, which cuts hs at R with an output of 2.9 million units and a price P_4. Thus the housing supply industry jumps from meeting only replacement demand (not shown in the diagram) in year 1 to an output of replacement demand plus 3.8 million units for stock in year 2, plus 2.9 million in year 3 and decreasing in subsequent years until a new equilibrium is established. The large fluctuations in demands on the industry which arise from an increase in demand for the services provided by a capital good are discussed in greater detail in Chapter 5.

Many different housing markets

At several stages in the foregoing analysis reference was made to the different types of housing demand and supply, notably housing to buy and housing to rent, housing provided by the private sector and that by the public sector, sometimes at a subsidised price. In practice there are separate and overlapping markets for a large range of types of housing, including flats, houses, bungalows; houses in various localities; as well as the more obvious rent/buy markets and the private-sector/public-sector choice. To consider the whole housing market as one is a simplification which must be dropped as the analysis is used to understand specific situations. Analysis, below, of the private-sector demand and supply position for housing to rent and housing to purchase, provides a framework within which other specialised demand and supply situations may be examined.

Housing to rent and housing to purchase in the private sector

The demand for housing to purchase and housing to rent is complex because the individual is likely to have a definite preference either for rented or for purchased accommodation, but will, if he has the financial resources, be prepared to switch from one to the other if the price of one becomes too high or if the supply is not available.

Consider the situation in Figure 4.6. RD_1 is the demand curve for housing units to rent made up of the individual demand curves whose first choice is to rent rather than buy. Curve PD_1 is the demand curve for all other persons, that is those whose first choice is to buy. RS is the stock of dwellings for renting, PS is the stock of dwellings for ownership.

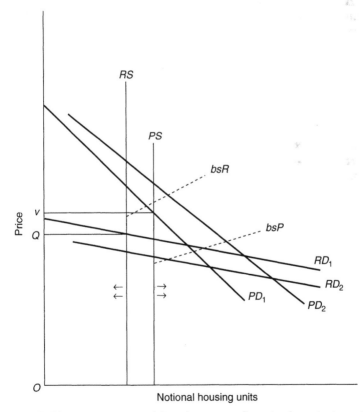

Figure 4.6 Housing to rent and housing to purchase in the private sector

In the rented sector, the market price for rented dwellings would be *OQ*, that is, the price at which supply curve *RS* (= stock) cuts the demand curve RD_1.

In conditions where there is some rent control and uncertainty as to its future extent, the housing industry (including developers) is unlikely to be confident in the long-run rate of return from housing to rent and will therefore be prepared to supply only at a rather high price, i.e. the building supply curve for rent is *bsR*. This lies above the demand curve and therefore no building will take place. For the same reason, there is unlikely to be appreciable net movement of housing from the stock for sale to the stock for rent.

In the market for purchase, the market price would be *OV*, the price where PD_1 cuts *PS* (the supply curve). The building industry supply curve *bsP* is such that the industry would be prepared to build and stock would eventually increase as in the case of the increase of the stock in Figure 4.5A–B.

There would not, however, be equilibrium because the demand curves PD_1 and RD_1 represent only first choice and there would be some reallocation once prices and availability are known from those wishing to rent to those wishing to buy, thus shifting the demand curve for purchase to, say, PD_2. The shape of the new demand curve will depend partly on the reasons for wishing to rent in the first place, e.g. whether for mobility or lack of capital (security) or income. The building supply curve *bsP* cuts the new demand curve PD_2 further to the right, thus increasing the output of the industry and the stock.

Corresponding to the shift to the right of the demand curve for purchase there will be a shift to the left of the demand curve to rent, to, say, RD_2. This would lower the market price and would attract back some persons from the purchase market, thus shifting that demand curve slightly to the left. At the same time as the demand curves are shifting, so will the supply curves move. As the price of housing to rent is lower than the price of housing to purchase, there will be a shift of accommodation from the stock for renting to the stock for purchase as indicated by the arrows in Figure 4.6. There will thus be a series of adjustments of decreasing size until equilibrium is achieved. These adjustments will be of three types, namely:

(*a*) changes in the choices of the consumers for accommodation to rent and accommodation to purchase;

(*b*) changes in the stock of accommodation to rent and to purchase as a result of a movement from one to the other;

(*c*) changes in the stock of accommodation as a result of new building.

Forecasting the demand for new housing

Although the theoretical analysis provides a background of fundamental understanding of the housing market, in practice only parts of it can yet be used as the basis of forecasting. There are broadly three methods of forecasting, all using aggregated data:

(*a*) to build up the total 'needs' for new housing and then to decide over what period need will be satisfied;

(*b*) to consider directly the factors which affect economic demand and forecast these to arrive at total housing demand;

(*c*) to look at past trends in housebuilding and in the factors which affected it and project these forward.

Need-based forecasts

Method (*a*) above uses assessment of 'need' for all sectors of housing together. Need is defined as the difference between an accepted standard of housing provision and the extent to which the stock reaches this standard. It is dependent on requirements for net new household formation; requirements to replace dwellings cleared in slum-clearance schemes and for other purposes such as road building; and requirements to increase the stock in relation to existing households to provide a margin of vacant dwellings for mobility. In addition, it is usually recognised that not all existing dwellings will be available to provide first homes as some will be used for second homes. This need assessment assumes that certain standards of housing provision are to be achieved. These standards are often laid down by government and therefore can change suddenly, but the underlying basis for them is usually public opinion or the social conscience, which tends slowly to raise its conception of what is an acceptable minimum as the standard of living rises.

The future number of households will, apart from immigration and emigration, be affected by the population and its age structure. Both these are known in so far as they affect the number of households, say, up to fifteen years ahead, because all heads of households are now born. Death rates are relatively constant. On the assumption that the same number of persons in any age group will head households (the headship rate) as at present, the total number of households can be predicted with reasonable accuracy.

This last assumption is, however, a very doubtful one. Firstly, the marriage rate may change and in any case unmarried partnership must also be considered; secondly, there is a tendency for more young, single people to leave the family unit and establish separate households; and thirdly, at the other extreme of the age range, widows and widowers are more prepared to live separately rather than live with the younger generation. Some definite forecast must therefore be made of headship rates.

In conceptual terms an increase in the number of households will add to the number of individual demand curves which have to be summed to arrive at the total demand curve. It will also shift the individual demand curves because of the different requirements in space and amenity of the traditional household with children living in, and of the new smaller households. Although one would not expect the total demand curve for housing units to go up in proportion to the number of households because the households are small, it would be expected to rise substantially because of the relatively high minimum requirement of a separate household. The space and amenities required per person are generally greater for a small household than for a large one.

Once the standard has been set, information on the existing dwellings is required to determine how much of the stock needs replacement. In the UK, census data on the condition of the housing stock are available at regular intervals. The major problem with this, and to a lesser extent other aspects of the need forecast, is to determine over what period of years needs can be met. For this purpose, recourse is usually had to method (c), that is, to a consideration of what the actual rate of construction could be in the foreseeable future.

The number of houses which should be replaced each year will in most cases be greater than would be demanded at the ruling price by the occupants of the slums, either because they do not have a high degree of satisfaction at the prospect of moving to a better house or because they have too low an income. The public replacement programme may be higher than that which would be achieved by the economic demand of individuals, and therefore social policy has to be translated into economic demand by some sort of public subsidy. Similarly, the vacancies desired for mobility may be more or less than would ultimately be provided by economic forces. Second homes by contrast are entirely a matter of demand (of individuals of firms) and not of need.

Forecasts of economic demand

The forecast of incomes to which to apply the elasticity of demand is in itself difficult, for, because of the underlying individual nature of demand, it is not sufficient simply to forecast the growth in personal disposable incomes. The distribution of these incomes is very significant and hence forecasts should be made, *inter alia*, of the level of unemployment and the relative rates of increase in wages and salaries and other types of income.

The supply of credit or savings must also be taken into account in forecasting economic demand. So far as the demand curve of the individual is concerned, if capital is expressed as income equivalent, no adjustment is required for credit, as credit is simply the capital which the income can command. So long as the rate of interest is holding the balance between the demand and supply of credit, no further theoretical adjustment is necessary. An increase in the supply of credit or savings, that is, capital, means a shift in the demand curve.

In order to predict in aggregated terms whether this shift will take place and by how much, changes in savings by private individuals are important and also the supply of credit. There may be situations where the loans are restricted by the inflow of funds and the demand for credit is not therefore being met at the ruling rate of interest. This means that the effective demand for housing is not represented by the total demand curve built up by summing individuals' demand curves on the method outlined, but is shifted downwards by the overall credit shortage. Changes in the rate of interest alter the price of housing in so far as this is represented by annual ongoings on house purchase.

Extension of past trends

This last method of forecasting is most useful when it is a reasonable assumption that there are no great changes in the underlying factors determining demand and supply for housing or when the period is so long that fluctuations around a trend are unimportant.

Attention is then concentrated on the changes which can reasonably be foreseen in the components of housing demand and the trend is adjusted as necessary. The factors which are considered in such forecasts are similar to those used under methods (*a*) and (*b*) but particularly those in (*b*), including incomes, supply of credit or savings, the rate of interest, house prices, etc.

United Kingdom Forecasts

Most regular short-term forecasting of economic demand for housing, notably that undertaken by the National Council of Building Material Producers and by Construction Forecasting and Research (CFR) Ltd., for example CFR (1999), is an amalgam of extension of past trends and a consideration of how changes in conditions are likely to affect those trends. Very little use is made of elasticities. Some forecasts of the economy as a whole, include housing as a category.

UK government organisations – Office of National Statistics (ONS) and Department of Environment, Transport and the Regions (DETR) – produce long-term estimates of number of households based on trends of population, marital and partnering projections, migration and so on (DETR, 1995). These are then examined by Regional Planning Conferences and ultimately the amount and distribution of housing requirements are determined for the whole of England. These forecasts are updated and reinterpreted from time to time, notably at the instigation of the Joseph Rowntree Foundation and other charitable housing organisations (Holmans, 1995; Holmans *et al.*, 1998; Holmans and Simpson, 1999).

5
Demand for Industrial and Commercial Building

The demand for factories and offices is not dependent directly on the ultimate consumer, but on those who produce goods for them. It is known as derived demand. Its analysis benefits from the concern of economists with the relationships between investment and consumption. The initial discussion will be mainly in terms of industrial building, followed by a consideration of the extent to which the same analysis is applicable to commercial building.

Acceleration principle

The acceleration principle states that if the output of any consumption good increases, the demand for the investment goods used in its production will increase at a greater rate. This may be illustrated by a numerical example, as shown in Table 5.1.

In year 1, year 2 and year 3 the output of good A is constant at 100 and the stock of buildings to produce A is also constant at 200. Five per cent of the buildings are replaced each year so that the demand from this source is constant at 10. Then in year 4 the demand for good A rises by 20 per cent to 120, and the demand for the stock of building rises to 240. Provided that the rise was anticipated, the new buildings to produce this output in year 4 should have been built in year 3. Thus in years 1 and 2 there is no demand for net increase in buildings. In year 3 there is a demand of 40 to enable production to be increased in year 4. In year 5 the demand for good A continues to rise, but at a decreasing rate, to 130. The demand for the stock of building rises to 260, showing a shortfall of 20 which must be constructed in year 4. Note that the demand for good A is still rising but the demand for new buildings

Table 5.1   Illustration of the acceleration principle

Year	Output of good A during year	Stock of buildings to produce A during year	Demand for new buildings to produce A to be built in preceding year		
			Replacement[a]	Net increase[b]	Total
1	100	200	10	0	10
2	100	200	10	0	10
3	100	200	10	40	50
4	120	240	12	20	32
5	130	260	13	0	13
6	130	260	13	−20	0
7	120	240	12	0	5
8	120	240	12	0	12
9	120	240	12	0	12

[a] Assuming replacement equal to 5 per cent of stock per annum.
[b] Assuming buildings constructed in year prior to increase in output of good.

is falling because it is related to the rate of increase in total demand which is already falling. By year 6 there is no longer any expansion in demand for good A and the demand for new building in year 5 falls to the replacement level of 13. If in year 7 the demand for good A decreases by 10, then the stock requirements will decrease by 20 and there will not even be a replacement demand in year 6 because 13 units of the surplus stock of buildings will be used for replacement. In year 7 there will still be 7 units of surplus capacity to set against the replacement need of 12, so that new building will be only 5 units. Not until year 8 will there be a real replacement need of 5 per cent of the stock of buildings.

The acceleration principle operates because buildings have a long life so that the stock constitutes a very important element in the total situation. If the life of buildings were the same as the production period under consideration – a year in this case – the demand for buildings would increase at the same rate as output, although a year in advance.

The acceleration principle therefore applies to all investment in buildings, as is clear from the housing analysis where, the stock of buildings being high, the new demand for buildings is very dependent on the increase in demand for housing. The only reasons why, traditionally, the acceleration principle is used to help explain the demand for industrial buildings rather than housing, are that (*a*) changes in demand of most manufactured goods for ultimate consumption are more obvious than changes in the demand for housing, and (*b*) the housing demand is so dependent on other determinants such as government social policy, which may mask the effects of the accelerator.

Practical modifications in application of the acceleration principle

In fact, fluctuations in the demand for industrial building do not follow the acceleration principle alone. Some of the reasons are as follows:

(*a*) There is often some surplus capacity in industry and it is therefore possible to increase output by increasing employment by overtime, shift working, etc.

(*b*) An industrial building is only a casing for the manufacturing process. The layout of machinery within a building can often be completely altered and modernised without altering the building itself. Even if the building is altered, this might appear in the statistics as repair and maintenance rather than new industrial building.

(*c*) If the increase in demand in manufacturing industry is not accompanied by expectations that the demand will continue, then output will be increased by other methods even though in the long run they might be more expensive. Thus, in Table 5.1, if the entrepreneur had anticipated that the increase in demand in years 5 and 6 would be temporary only, he would not have increased his stock of buildings.

(*d*) The willingness of entrepreneurs to increase their capacity will depend, in addition to expectations of demand, on profits and expectations of profit. Thus, if demand is rising and costs are expected to rise too, then expectations of profits may be such that entrepreneurs will not expand. Similarly, their ability and willingness to expand will depend on other factors such as the availability of capital and the cost of capital.

(*e*) Ability to expand may be affected by government policy, in the price and availability of credit, by physical controls on development, for example through planning regulations, and also by the involvement of governments in the provision of industrial building, often through a statutory agency.

(*f*) Technological change may indicate requirements for new buildings for replacement: for example, improved atmospheric conditions required in cotton manufacture may need new factory buildings and the increase in services in office buildings require large cavities between floors. The rate of technological change and hence of the increase in demand will fluctuate over time.

Response of demand to price

It is important for the construction industry to have some idea of the elasticity of demand for industrial buildings. Because the demand is derived, it is dependent on many more things than price. For example, an industrialist who is considering whether or not to increase his productive capacity by putting up a new factory must consider the overall profitability of the capital invested and will probably not be willing to undertake the project unless he has an expectation of a high return on capital for many years. The capital invested may be, say, 20 per cent buildings and 80 per cent plant and machinery, and if he wishes to undertake the project he must have both together. Consequently, a rise in the price of building by 10 per cent will be a rise in the cost of the project by 2 per cent. This is very small in relation to the margin of error in the calculation of his expected profit. Moreover, the capital costs will in many cases be a relatively small part of total costs. Hence, the demand curve for industrial buildings is likely to be relatively inelastic.

Forecasts of the level of industrial building

Because of the close relationship between the level of industrial investment, the level of industrial production and hence the health of the economy, it would be expected that forecasts of the level of industrial building could be made on the basis of certain assumptions on the level of industrial production and gross domestic product. Various attempts have been made to use econometric methods to obtain long-term forecasts, for example by the Joint Working Party on Demand and Output Forecasts of the EDCs for Building and Civil Engineering (1971) and the continuing models of the total economy give broad indications of industrial investment.

The most obvious relationship is that given by the incremental capital–output ratios (ICORs). This is the relation between net capital formation in any period and the increase in output during that period. In the example of Table 5.1 above, because the output has been lagged a year after investment the ICOR has an odd value of 40/0 or x in year 3. Had the output not been lagged, the ICOR in year 4 would have been 40/20 or 2 and in year 5 20/10 or 2. In all other year except year 7, when it would be negative, the ICOR would be nil. Various adjustments may be made to this crude ICOR, for example to exclude changes in output attributable to employment rather than to capital. The method depends on a reliable forecast of output, and once a figure for investment has been obtained this still has to be apportioned between plant and equipment and buildings and works.

Because of the apparently closer relationship of investment in plant and machinery than investment generally to output of manufacturing industry, attempts have been made first to forecast investment in plant and then to relate this to the output of buildings. This is essentially using the same relationships but undertaking the analysis in a different order and with perhaps a more logical sequence.

Other methods which depend less on forecasts of exogenous variables include relating investment in factory buildings directly to lagged gross domestic product and to profits, also suitably adjusted for the time trend.

None of these various methods has been found to give results which, when applied to periods in the past for which the outcome is known, inspire confidence in their use as the only method of forecasting. In the last resort, forecasters have to make a judgement of the likely level of output based partly on the results of econometric analysis but mainly on past trends and the factors likely to affect them. Forecasts of the national economy sometimes specify investment by manufacturing industry and this is sometimes broken down but rarely split between building and plant and machinery.

Commercial building

By commercial building is meant the whole range of buildings other than industrial buildings for an enterprise engaged in the buying or selling of a product or service. It thus includes offices, hotels, garages, shops and so on; that is, nearly all private building except for housing, industrial building and a few miscellaneous categories such as churches. It may or may not include some projects which may be provided privately but with public sector participation and control: for example, water supply and power. Whether the economic analysis of commercial building is relevant depends on the individual circumstances of demand determination.

There is, on the face of it, no reason why the investment in commercial buildings should not be related to output, growth, profits, etc., in a similar way to that for industrial building. The basic characteristic of long life in relation to the services or goods it helps to produce is present and so is a relationship to productive capacity.

This area could, therefore, be subjected to the same kind of research for econometric relationships as industrial building. The problem is that some of the difficulties mentioned for the application of the acceleration principle to industrial building apply even more to commercial building and, in addition, there are others:

(*a*) Commercial building is a more heterogeneous collection of buildings and therefore different equations for each type of building would need to be established. Data are not available in sufficient detail to enable this to be done. Moreover, as the types of building are more distinct (compared with the totality of industrial building), it becomes less relevant to consider their relationship to general factors such as output. The individual variations would not be expected to cancel each other out. More specific factors likely to affect types of commercial building include changes in tastes for entertainment, and technological change, for example, automatic garaging.

(*b*) The capacity of an office and of some other types of building is less rigid than that of factories. It is usually possible in the short run to squeeze in extra personnel and equipment by lowering space standards.

(*c*) The importance of expectations of output and of profits, and of the availability and cost of financing capital, are as important for commercial building as for industrial building, but the users of commercial buildings and the financiers of them are often different people, thus complicating the demand factors.

(*d*) The state of the economy and of government policy is similarly important. For example, government has affected development in selected urban areas of Great Britain.

(*e*) The commercial sector is probably more concerned to present a good public image than the industrial sector and may be dependent on this for its profits, for example, in retailing.

As in the case of industrial building, the elasticity of demand for commercial building is likely to be low. However, because of (*b*) above it may be rather more elastic than that of industrial building.

Forecasts of commercial building are usually based on a qualitative analysis of the underlying factors affecting the various sectors, sometimes but rarely supported by econometric work along the lines of that mentioned above under industrial building.

Other derived demand

Although the clear cases of derived demand are industrial and commercial building in the private sector, in some countries much of public-sector demand is substantially derived demand. Clear examples are the nationalised electricity and gas industries. Others, such as transport and water supply, are partly derived demand and partly determined by social policy.

6
Demand for Social-type Construction

The two preceding chapters on demand for construction, started by considering the factors which were relevant to the decision to build in a broad sector of demand, and the theoretical concepts which were of assistance in understanding the interrelationships of the relevant factors. This chapter deals with a heterogeneous sector of construction in which the common features are that the product is used by a large number of persons or households collectively – for example, hospitals, museums, roads, schools – or that it is used by persons or households individually, but who individually are neither able nor willing to pay for the product, but which the 'community' decides should be available.

Housing in the public and quasi-public sector comes partly into this category, although it is also dealt with in Chapter 4 on demand for buildings for the direct enjoyment of individuals. In that chapter the supply curve of the housing supply industry was taken as given, and consideration was focused on the demand side. The supply curve for social housing is, however, determined by the same sort of considerations as those affecting other types of social demand and hence social housing is also included in this chapter.

Difficulties of a theoretical approach

Welfare economics can make substantial contributions to decision-making on many matters of public expenditure, but on choices between comparatively small areas of expenditure all widely different in kind it has little to offer. One reason for this is the lack of any basis for making interpersonal comparisons of satisfaction, so that the indifference curves of individuals between various types of construction expenditure cannot be summed. One is inevitably led to a sort of 'superman' (Little, 1957, p. 121) deciding what is good for the community, and a theory based on the opinion of one individual is hardly better than no theory at all.

Cost–benefit analysis is a technique which enables some of welfare economics to be applied in a practical situation. It rests on one of the basic theses of welfare economics that, if the sum of the value of the benefits of a course of action is greater than the sum of the costs of that action, then it would be possible in theory for those who receive the benefits to give compensation to the losers and still be better off. In this practical application there are great difficulties in assessing the value of advantages and disadvantages of a particular course of action. Thus, its principal use tends to be in the comparison of the relative net benefits of similar alternative courses of action. Hence, where the errors in the method of assessment of costs and benefits occur to roughly the same degree in each alternative under consideration, their absolute magnitude is less important. Thus it is feasible to compare sites for an airport because the *types* of advantages and disadvantages are likely to be the same. However, the technique has much less validity when comparing, say, investment in hospitals with investment in airports, because one is concerned with, for example, putting a value on life and on health and comparing this with the value of greater business facilities and holidays. Clearly, the margin of error or disagreement in the assignment of such values is so great that it may well swamp the differences between the final assessment of net benefits or losses.

Prest and Turvey (1965, p. 683) came to a similar conclusion when they said that:

> The technique is more useful in the public utility area than in the social services area of government. Comparisons between, say, different road projects are more helpful than those between, say, road and water projects; and both these are likely to be more helpful than application in the fields of education, health, research and so on.

Forecasting in practice

Notwithstanding the lack of theoretical help, it is important for the industry to be able to make some assessment of the social-type demands likely to be put to it. Of the three basic methods of fore-casting outlined in Chapter 4, the most relevant to this sector are an assessment of needs together with a consideration of the re-sources likely to be available.

Assessments of standards

Standards of provision of services are not clearly defined, although some implicit standards usually exist and can be taken as a foun-dation on which to build. The very discussion of what minimum standards ought to be is helpful in policy-making. In housing there are two levels of standards: the level below which no dwelling ought to be occupied, that is, slums; and the standard below which no new building should ideally be constructed. Between these two levels is an enormous range of conditions, and the lower-level standard may at any time be raised to include another chunk of housing. In studying demand, however, so long as the lowest level of standard is not achievable at any practicable rate of construction for many years, it is not necessary to consider the area in between the two standards in great detail because it is unlikely substantially to affect current policy. This is unfortunately the situation in most develop-ing countries.

The same applies to other areas of demand. If the lowest possible minimum standard is clearly not being achieved, then to argue at length on the acceptable minimum standard above this lowest minimum level will often be quite irrelevant for policy formation. Thus, in the case of pollution, it is not fruitful for demand assess-ment to discuss the standards which should be established, when the first priority, and one which would occupy the authorities for a number of years, may be to attain the lower standards set by existing legislation. The first step, therefore, is to assess the lowest minimum standard and see in broad terms how nearly it is reached. Only if it is approached by present provision is it necessary to study in detail the changes taking place in policy and opinion which will establish a new standard.

Size, age and condition of the stock

Once some assessment of a standard has been made, it should be possible to assess the present stock and consider how far it falls short of the established standard. Whereas the standard itself is a matter of opinion and may therefore be difficult to assess, the stock is a matter of fact and is not basically difficult to establish, although the resources required may be substantial. For most of the public, socially determined sectors, very few data are available, with the possible exception of housing, schools and roads. Without these data any study of demand becomes very dubious. However, if the information can be obtained, some assessment of present need can be made.

Future increases in need

Once present need is established a further step must be taken, namely the assessment of how far this need is likely to increase in the future, either because of changed standards already discussed, or changed populations to which these standards apply. Thus, if the population of households increases, the number of dwelling units required will increase. If water provision is partly related to industrial output and industrial output increases, the consumption of water will increase.

Time period of provision

Having established present and future need, it is still necessary to determine at what rate this backlog can be diminished. In the case of housing, the wide variations in total demand assessments are principally accounted for by the different assumptions on the rate at which need will be transformed into effective demand. This is an area in which political decisions and priorities play a great part. However, the problem can often be narrowed by consideration of the constraints in transferring need into effective demand.

Constraints

The first constraint (if it can be called this) is that in the bargaining process which goes on between protagonists of various sectors; once a sector has achieved a high rate of growth for a number of years it may well become the 'turn' of another sector. For similar reasons (as well as those discussed before) it is probably difficult for any sector to grow at an extremely high rate unless there is a great pressure of public opinion in its favour – as there may be, for example, for measures against pollution in Western developed countries or for low cost housing in South Africa.

Apart from the political difficulty of a sector growing rapidly, there are practical difficulties. In the case of slum clearance, for example, to increase the rate of replacement in, say, five years' time may require action now on demolition orders, re-housing, clearing of sites, etc., and there may well be administrative and managerial bottlenecks even if there are no financial problems. To increase the rate of slum replacement in one year (assuming these preparatory processes have not been undertaken) may be virtually impossible.

There are three stages in the construction process. The first is the conceptual phase when, for example, planning the route of a road must take place followed by public enquiries and land acquisition. For roads and harbours and commercial building this phase can be up to ten years but for other types of work two to four years is not unusual. Then there is the design and contract documentation phases which range from six months to about four years. The construction may then take months or years.

Lastly, if the rate of increase in this socially determined sector corresponds to a rapid rate of increase in other sectors, there could be inadequate capacity in the construction industry itself or in some of the materials industries. In 1964, 1973 and 1989 in the UK the construction industry was stretched to capacity and there have been shortages of materials from time to time such as plasterboard, bricks, copper pipes and even manhole covers. Cladding was imported in the late 1980s.

Response of demand to price

Formerly in the UK, when the system of cost limits for individual building types operated in large parts of the public sector, the overall cost of a unit of output was fixed, that is, if in school building, for example, the cost of building a place rose by 1 per cent, then to keep within the cost limits, standards would need to fall by 1 per cent. In this case the elasticity of demand is 1.

In general, various control systems on public spending make the relationship for any one project to prices less direct, but the elasticity of demand for most social goods is likely to be inelastic.

In the case of individual projects demand may be very inelastic. Examples are the Royal Opera House in Covent Garden, UK and the Channel Tunnel. In both cases costs went up, with a great increase in total cost, but the work done altered little. There must be a number of other large projects where, partly because of the high real costs of abandoning work part of the way through a project, the costs have increased but have caused no diminution in the amount of building produced; that is, there is a very low elasticity of demand.

7
Demand for Refurbishment, Repair and Maintenance

Definitions

The demand for modernisation, rehabilitation and refurbishment of buildings and works is dependent on factors similar to those relevant for the equivalent type of product of a new build. Aikivuori (1996) describes refurbishment as being an option at the end of the service life of a building and this occurs when the building fails to perform as required in use. She identifies three causes for this failure: deterioration in the building, changes in the requirements of performance of the building (obsolescence) and change in use. Maintenance is described by Wall (1993) as encompassing 'that ongoing process dealing with maintenance or restoring to good condition any part of a building that becomes defective or non-functioning to an acceptable standard to sustain the utility and life of a building (Building Maintenance Committee, 1972; Dixon, 1990; Mole (1991)'.

Repair and maintenance are also very different from rehabilitation for the following main reasons:
- they are normally paid for out of income rather than capital;
- although some of the work is postponable for a period, if it is postponed, the requirement for expenditure is likely to increase owing to the neglect of basic work.

The needs for refurbishment and for repair and maintenance are both due to deterioration of the building. The causes of deterioration are different although overlapping. One may be the expected lifespan of basic structural materials which cannot be substantially altered by maintenance, for example, ageing of slate roofs or bricks, and another may be a deterioration of materials such as paint or timber which can be replaced or whose lifespan can be considerably extended by regular attention. It is the latter which is the subject of repair and maintenance. Repair takes place after some problem has been diagnosed. Maintenance is concerned more with prevention.

Factors affecting level of refurbishment

The level of refurbishment work depends on the cost of refurbishment compared with the cost of building new. Decisions will then be taken along the same lines as for new build. As with new building, this type of work is financed from capital rather than income and it is often postponable. For housing, the individual's and the economy's demand will depend on preference for improved housing compared with all other goods and on its price (see Chapter 4). For industrial and commercial work, consideration will be given to the demand for the ultimate product the buildings help to produce and hence will be dependent on the state of the economy and on expectations (see Chapter 5). Similarly, refurbishment of social-type construction products depends on the need for such improvement and the competing claims on resources, especially finance. It also depends on the state of the economy and the current views on how it should be managed and hence the level of permitted public sector or quasi-public sector expenditure (see Chapter 3).

The factors affecting the demand for refurbishment are not purely economic ones, however. In many parts of the world there are concerns to protect the cultural heritage, whether this be substantial grand buildings and civil engineering works or traditional housing. There is even greater consideration given to buildings of special architectural interest. Thus, many buildings are being conserved even if it is more costly to do so than to knock down and build new. This trend towards preservation is bolstered by the wish that non-renewable materials should not be wasted for environmental reasons.

Factors affecting levels of repair and maintenance

In an ideal society, repair and maintenance would be undertaken regularly at intervals for each component part of the building, as appropriate for its continued existence in good condition. In practice, this frequently does not happen and maintenance as well as repair is often postponed. Since repair and maintenance are largely paid for out of income, one of the major reasons for maintenance not being undertaken is a shortage of income. It would be expected that income elasticity of demand for repair and maintenance would generally be high, but this would not apply evenly through the economy. The willingness to spend on repair and maintenance is affected by the stake that the persons responsible for repairs have in the building. Owners are concerned to safeguard their capital. Occupiers of property are concerned to make their occupancy comfortable or profitable. If owners are also occupiers, the reasons to undertake repair and maintenance are multiplied. For this reason it would be expected that owner-occupied housing would have a particularly high income elasticity of demand for maintenance.

As it happens, it is possible to do a very basic check on the income elasticity of demand for housing of all tenure types in the UK, using the Family Expenditure Survey (ONS, 1998a). The survey looks at the expenditure of 6,409 households divided into 10 income groups. Of these 67 per cent are owner-occupiers and the remainder rent. Nearly all of the three highest income groups own their properties, compared with under half of the three lowest income groups. Disregarding the top and bottom income groups, and taking the mean of each income range, the percentage change in income from top to bottom is 6.8. The percentage increase in expenditure on repairs, maintenance and decoration is 5.3. This gives an income elasticity of demand of 5.3 divided by 6.8, which is 0.8, showing that income levels and/or tenure have an important influence on the amount spent.

At national level it would be expected that the amount spent on repair and maintenance would be related, first, to size of the stock of buildings and works and to its age composition and, second, to income level: GDP or GDP per capita. For the UK there are data on output of repair and maintenance (though for Great Britain only and including some refurbishment) (DETR, 1998b) and on GDP and GDP per capita (ONS, 1998b), though information on the size and age of stock is scanty. Unfortunately, it is necessary to use data at current prices for both GDP and repair and maintenance because the price correction factors for the latter are believed to be unreliable. The increase in GDP per capita from 1989 to 1997 at current prices was 52 per cent. The increase in total repair and maintenance output was 24 per cent and in all housing repair and maintenance 20 per cent. This gives income elasticity of demand for repair and maintenance at the national level calculated over time of 0.5 for all repair and maintenance and 0.4 for housing repair and maintenance.

It would be interesting to look at the effect of differences in GDP, from one country to another, on the amount of repair and maintenance undertaken but, unfortunately, there are no reliable statistics on the amount of repair and maintenance in poor countries and data, even for developed countries, are dubious. In spite of poor statistics, however, there seems little doubt that the amount spent on repair and maintenance varies greatly between developed and developing countries and between market and former planned economies. In the UK, for example, it is around 50 per cent of all construction, whereas in most developing countries such data as are available suggest that it is nearer 10 per cent. Indeed, it is only necessary to observe the state of the stock of buildings, roads and so on in many developing countries to see that repair and maintenance are grossly inadequate. There are several possible reasons for this apart from low incomes:

- Finance for government-funded projects is very short and priority is given to creating new assets rather than maintaining existing assets. This is partly because there is likely to be greater political advantage in being seen to initiate schemes, especially if they have popular appeal and are suitable for public inauguration.
- Once buildings have fallen into disrepair from lack of maintenance, it is often seen as cheaper to demolish them and build new.
- Much of the stock of large buildings and infrastructure has been financed by international or bilateral aid or loans. However, the finance rarely covers provision for repair and maintenance expenditure.
- Because of the considerable external finance for buildings and infrastructure, the stock of buildings and works is considerably larger than could have been financed internally, and the requirement for repair and maintenance expenditure is far in excess of what the GDP can support.
- There is little incentive for bilateral donors to finance maintenance because, whereas in giving aid tied to their own contractors for new projects they are helping their own industry, for repair and maintenance their own contractors would not be interested in the work.
- The technology used for these externally financed projects is often sophisticated Western technology, such as high-rise reinforced concrete buildings, for which traditional maintenance methods are unsuited.

Repair and maintenance in developing countries may be seriously undercounted, as indeed is the new construction of very simple dwellings. It is possible that, for simple housing built with traditional materials, repair and maintenance is relatively higher, both in relation to the initial cost of housing and in relation to incomes, than for many other types of structures. This would be because, although the housing is very basic, it would have high priority for householders. Even more important, however, is the fact that repair can be done by the householder with easily available materials and techniques, since he would often have built the house himself with help from family and friends. Regular maintenance might even be planned; for example, in Tanzania reeds were gathered and dried several weeks or months ahead of the rainy season, ready to repair roofs later.

In many of the former Soviet Union (FSU) countries and their satellites, repair and maintenance was not undertaken on a regular basis, or a least not on a scale which was adequate. Here the reasons must be different. It is difficult to understand that the Soviet planners did not realise the vital importance to the long-term health of the economy of keeping buildings and infrastructure in good working order, but it seems either that they did not or that the pressures to build new were so great that it was politically impossible to allocate resources to repair and maintenance. However, the strength of political pressures to build new and to fulfil the requirements for new projects which were seen to be evidence of development and had the important characteristic of 'inaugurability' was probably the main reason. When there has been a low maintenance economy for a long time, difficulties arise in reversing the trend. Much of the new construction was high-rise, of reinforced concrete panels which require fewer traditional skills. After some years the supply of craftsmen able to maintain older buildings was not there. Moreover, some of the technology of the new building makes it necessary to have a major refurbishment after 25 or 30 years which necessitates evacuation of the factory or dwellings – again a politically difficult decision. Another reason for the problems in the FSU is that the owner and the user of buildings were not the same. Where housing is owner-occupied the situation in regard to repair and maintenance is generally better.

8
How Demand is Put to the Industry

In order to appreciate the reaction of the industry to the demands upon it, it is necessary to understand the way in which demand is placed. In this chapter a brief description of the construction process is given with emphasis on those aspects which affect the type of theoretical analysis required to understand the behaviour of construction firms.

Clients

The initiators of the whole process are the clients of the industry. The traditional first categorisation of clients is into public sector clients and private sector clients. This division is based on the person or organisation paying for the project rather than the type of project, for example social-type projects, as described in Chapter 6. To the classification according to who is paying, it is important to add, firstly, hybrid projects paid for entirely or partly by private funds but supplying social type goods and, secondly, projects financed from international aid.

This description of projects based on who pays is very important, but it is arguable that in many countries perhaps the more important subdivisions are, firstly, whether the client is experienced and knowledgeable or a one-off client and unsure of how the process works and, secondly, whether the project he is commissioning is large and complex or, at the other extreme, small and simple.

Two types of private client must be considered. The first is the client commissioning work for his own use, for example a company building a factory or an individual commissioning the design and construction of his own house. The second is a developer who is a client of the industry because he wants to sell or let the finished building. The latter may be divided into those who are also builders and undertake the function of the main contractor themselves and developers who obtain the services of a contractor to undertake the work. In countries with a developed capital market, the developer has a major role. In the UK for instance, there is considerable activity by developers in commercial and industrial work. In addition, those who participate in PFI projects are a new form of developer. In the UK most housing is built by developers.

There are complications in understanding the objectives of private clients in the industry because the user, developer and financier may be quite different persons or organisations. The initiator of the project and the client may be any one of these or occasionally in effect a fusion of them. Generally, they all seek value for money, but the user may have different preferences for quality as opposed to price than the others.

Partly because of the political swing favouring the operation of the market economy, the proportion of work in the private sector has grown. There certainly have been very great changes in the type and behaviour of private clients in the UK and, it is believed, to a lesser extent in other countries too. Change in the UK originated in the private sector and gathered momentum in the late 1970s and 1980s, largely led by clients for commercial buildings. Raftery (1991, p.131) identifies the key features of the changes as 'a reduction in the number of interfaces between the client and the industry, a clarification of responsibilities of members of the building team, increased attention to the *control* of time and cost, an explicit separation of the role of 'management' and a consequent reduction in the power and role of the architect.'

In the 1990s there has been a continued pressure to improve the construction process, initiated with the Latham report (Latham, 1994), and followed by the more recent Egan Report (DETR, 1998a). The basis of much of the thinking, especially in the Egan Report, is that it is possible and desirable to apply some of the lean production techniques of the Japanese car industry to the UK construction industry. Green (1999) believes that the arguments have ignored many institutional factors and human costs of lean production and is concerned that the move towards this approach is driven by the vested interests of large commercial clients. In addition, comment in the construction industry on Latham and Egan has tended to ignore several salient points which it is worth mentioning here.

The first is that the improvements in construction costs to which the documents refer are really changes in price, that is, the amount of money that the client pays for the project. The contractors costs are mentioned less. The second point is that the push towards an improvement in the service, which the client wants, took place in what was probably the worst recession in the UK construction industry since the 1930s, so that much of the reduction in price or improved contractual terms claimed to have been achieved as a result of these reports may have been due to the existence of a buyers' market, which forced a fall in contractors' profits and possibly damaged the long-run health of the industry (Hillebrandt *et al.*, 1995). Thirdly, although both reports speak of the industry as though they were dealing with the totality, in fact much of the discussion and most of the recommendations are applicable only to that part of the industry which is constructing large new projects, say to 25 per cent of the whole construction industry output. Nevertheless, these two reports and the related documents prepared by numerous working groups will have a substantial influence on the way large projects, especially those commissioned by large repeat clients, are put to the industry.

The client for small, new private sector work is often building for the first time and is therefore not knowledgeable about the construction process. His bargaining power in relation to the construction team is slight and, to obtain the appropriate contractor, he must rely on professional advisers, on competition amongst the contractors, on recommendations or on all three. For repair and maintenance work there is a greater chance that the client is experienced, since the need for repair arises relatively frequently. If he is not experienced he may lay himself open to poor service because the very smallest builders may not be capable of doing good work and a small minority may be seeking opportunities for dishonest trading.

Public sector clients fall into three categories: firstly, public authorities run more or less as commercial or industrial enterprises, although they may be subsidised by government; secondly, those providing infrastructure projects and other community goods used by individuals in the community but not paid for by the users; and, thirdly, clients who provide a service partially paid for by users, but not on a commercial basis. The importance of projects financed by the public sector has fallen in the UK and in many other countries, partly because of a disenchantment with the performance of government and quasi-government organisations compared to those in the private sector, partly because of the increased burden on the economy of public spending and partly because of the largely political switch to operation through a market process, referred to earlier in Chapter 3. With the possible exception of the first category of client, public sector clients tend to organise projects according to the rules laid down and to follow customary procedures. These procedures have, in the past, been dominated by the need to be, and to be seen to be, impartial in awarding contracts, even at some cost to efficiency. Thus, although they have for a long time, been large, continuing, knowledgeable clients, they did not use their power, except to a very limited extent, to try to obtain a better deal from the contractor. Changes did take place, for example, the decline in the use of open tendering, but the great push for change did not emerge until the 1990s largely prompted by Latham and Egan.

Some of the projects, formerly totally within the orbit of the public sector, are being undertaken by private organisations, such as the privatised water companies in the UK; for others the finance is provided jointly by private and public funds or entirely by private funds. The principal mechanism for arranging this in the UK is known as the Private Finance Initiative (PFI). Other countries, including developing countries, are adopting similar policies or considering doing so.

The clients for aid-financed capital construction projects may be an international organisation, such as the World Bank, or an individual donor country. If the latter, the client is likely to behave as is the custom in that country, with the contract being open to the contractors of the donor country. International organisations have their own procedures, usually with a pre-qualification stage and then a selected tender, often with far too many invited tenderers.

As a result of these developments, many new types of procurement have been developed or come into more general use. These are described in the sections below dealing with the roles of the participants and the selection process.

Procurement: the roles of the participants

The process by which the client, having decided to erect a building or undertake a civil engineering project, ensures that it will get built varies according to the traditional approach of the country, the size and type of work, the type of client and his perceived needs.

The traditional method, and that believed to be most widely used across most areas of the world, is described below. First, the client appoints his principal professional adviser and/or designer. In many countries he will automatically be a civil engineer since, in large parts of the world, the civil engineer has recognition of his status exceeding that of the architect. In the UK the professional adviser/ designer will traditionally be an architect for building and a consulting engineer for civil engineering work. Exceptionally, he could be a structural engineer or even a services engineer, depending on the nature of the project. In the public sector in many countries the principal designer may be within the client's own organisation and a few private organisations also have their own designers. The principal designer will then bring in other specialists to help in the design of the project. After the design is complete contractor selection takes place. Thus, in this system, the contractor has no part in the design process.

Non-traditional methods are being used in the UK widely in the private sector for large contracts. Meanwhile, for the public sector the government has turned a full somersault. While, at one time, it was hesitant even about abandoning open tendering, now it has issued guidelines to government departments encouraging the use of design and build and prime contracting (still barely tested or defined, see below) both of which give the contractor the lead role. It is also saying that traditional procurement should be used only if there are very good reasons in terms of value for money.

There have always been variations in the traditional process, but now in the UK the variants, so far as large projects are concerned, may be becoming more important than the norm. They may be divided into two groups. The first is those in which the contractor is selling a building or other facilities, often together with some other services, such as maintenance of the building or even its management. The second is where the contractor alone or in some form of partnership is selling a service, which includes the ownership and financing of whatever assets are required to provide that service. Some of these non-traditional arrangements are briefly described below.

Provision of a building or other facility

Project management

A project manager undertakes 'the overall planning, control and coordination of a project from inception to completion aimed at meeting a client's requirements and ensuring completion on time within cost and to required quality standards' (CIOB, 1982). The project manager may be within the client's own organisation or be a professional firm. Essentially, the project manager acts on behalf of the client and must decide how he is going to place the project with the industry. Project management may, therefore, be combined with a variety of different procurement practices. It is used extensively for large projects.

Design and build (D&B)

Under this system, the contractor is appointed to be responsible for design and construction of the whole project. Variants of this system have been used ever since the 1950s, often known as turnkey projects or package deals. There are two types of design and build arrangements. In the first, the contractor employs his own in-house architect or appoints a firm of architects to work under his direction. The contractor is truly in control. In the second, the contractor takes over the initial design commissioned by the client and employs the architect who has already been selected. The latter is known as novated design and build. It may be little more than a ploy to give the contractor liability for the work of the client's designers and as such is really general contracting with an additional risk. The advantages of true design and build are that it simplifies the chain of command for the client, who has to deal only with the contractor; it enables the contractor to contribute to the design at an early stage and therefore should assist 'buildability'; and it enables a saving of time (Hillebrandt *et al.*, 1995).

Management contracting

Management contracting is the arrangement whereby the whole process is managed by the contractor's team on a fee basis, with the management contractor having a contractual relationship with the subcontractors. This method of working enables the management contractor to be brought in at an early stage, while still enabling contracts to be let on a competitive basis. It was strongly favoured in the UK in the 1980s, but the number of contracts substantially declined in the 1990s (DL&E, 1999).

Construction management

Construction Management is the system whereby the role of management and coordination of construction is taken on by a professional firm, acting as a consultant to the client. The construction manager arranges trade and general building contracts direct with the client so that the latter has direct contractual relationships with the specialist and trade contractors. It sometimes occurs that the construction manager is also the designer. This system is similar to the old Scottish separate trades system, to the system adopted in France for smaller contracts and to the practice in a number of other countries. Construction management in the UK has decreased so that it is no longer of great significance (DL&E, 1999).

Prime Contracting

This is a system developed by the Ministry of Defence and others. Although two pilot projects are under way, the system is not yet in operation though it is already recommended by the Treasury. It has only one point of contact to the industry and uses a sophisticated planning system which aims to deliver the project ready to operate exactly as required, with no unnecessary attributes. It is intended to incorporate optimal through-life cost. It aims to give participants, including subcontractors and suppliers involved as part of a team, a satisfactory level of profit through a target cost–incentive arrangement. It achieves savings by better planning at all levels, rather than cutting margins. (Holti *et al.*, 1999a, 1999b). The method of selecting the contractor and determining his remuneration had not yet been decided at the time of writing.

Build, operate, transfer (BOT)

Under this system the contractor builds (and possibly designs) the project, then manages the running of the facility for remuneration based on the expected running costs. At a later stage the project may be handed back to the original client. This type of arrangement has been used, for example, for prisons and for health-care buildings.

In all these variants on the process, the important questions are whether the contractor has direct contractual responsibility for the work of trade contractors and whether the contractor has overall design responsibility for the work.

Provision of a service for the client

Build, own, operate, transfer (BOOT)

In this arrangement the client is being provided with a service which happens to require a building or other works to enable the service to be provided. The supplier is a service provider and may be a contractor, any other participant to the process or some organisation specialising in putting together such deals. It is risky for the contractor and his collaborators because costs of developing a proposal are very high but also because it is extremely difficult to assess the cost of, say, running a hospital over a period of several years. This method of operation is favoured by international development agencies and has been used, for example, for power stations.

Private finance initiative (PFI)

Since its launch in 1992, the UK government has developed the Private Finance Initiative under which contractors, usually in a consortium, are asked to bid in competition to finance, build and operate a facility, traditionally provided by the public sector, transferring to Government ownership at the end of a period of years. It is, in effect, a further development of BOOT, but the process becomes further complicated. One of government's objectives is to divest itself of risk and to reduce its capital spending. The danger is that, as the service providers are paid at regular intervals for the provision of the service, the current expenditure of government will, after some years, escalate to unsustainable levels. The risks to contractors increase substantially with this system of operation. Moreover, the costs of tendering for such projects – a type of transaction cost – are very high which means that all the parties to the tender need more working capital than in usual contracting operations. As at 1999, most of the projects have been large, but attempts are being made to extend the method of operating to smaller projects and smaller contractors (CIC, 1998). Negotiations continue about the precise conditions which should apply to PFI projects.

The diversity of practice for large projects and some smaller projects in the UK construction industry is evident from the above list. In fact, these classifications above are broad indications of practice only. There seem to be an almost infinite number of variants in procurement practice. The changes which have taken place in the UK may be summarised as an increase in diversity, an increasing role being adopted by the client or his representative, a desire for a single point of client contact and a wider scope for the contractor in providing services other than construction. Similar developments to those in the UK are also taking place in other countries, both developed and developing, and several developing countries are using or trying to use private finance for public projects on a BOOT basis. Uganda has used private finance for public market booths and is in process of arranging it for major infrastructure projects but these have not yet been finalised (Mugume, 1999). Japan is reported to be considering a form of PFI (Cavill, 1999).

No numerical information on the usage of these methods has been found but there are data on types of contract documentation used, gathered from firms of quantity surveyors in the UK. They refer to the larger projects only. They exclude speculative housing construction which is all design and build. This survey suggests that in 1998 about 20 per cent of contracts let were design and build, accounting for 40 per cent the value of contracts. Construction management and management contracting accounted for about 17 per cent of the value of projects (DL&E, 1999).

Procurement: method of selection

There are several ways of appointing the contractor in the traditional system. The most usual way is to invite tenders based on documents which may include drawings, a specification or, notably in the UK and areas of past British influence, a bill of quantities; that is, a document listing all the items in the building drawn up by the quantity surveyor from the available plans and drawings. For civil engineering works the consulting engineer usually performs the quantity surveying function.

The contractor, having decided to put in a tender, arrives at a price which will cover his estimate of the cost of undertaking the work, a contribution to overheads and his profit. He must make appropriate allowance for risk.

In the case of open tendering, the contract is advertised and any contractor may ask for tender documents and put in his price for the job. In selective tendering, the client or his adviser draws up a short list of contractors whom he would be happy to have undertaking the work, and each of these is asked to put in a price.

There are objections to open tendering, notably lack of control of the client over the competence of the builder he is employing and waste of resources when many firms tender for the same job, for the prices for the contract obtained must overall cover the cost of estimating on jobs not obtained. However, it does have the merit that it is seen to be fair and it is therefore widely used in developing countries for smaller public sector work. The Chinese have introduced open tendering in the last few years, as an alternative to a system of direct orders. In the UK open tendering is now virtually unknown. Much of the public sector work which was let by open tender is probably now let on selective tender or even on some less traditional way. Many countries have a contractor registration scheme which classifies contractors according to their technical and financial competence to undertake work of various types and sizes. The system acts as a sort of permanent pre-qualification system. If contractors are invited to tender from one or more of these groups, on a selective or open basis, they should be capable of doing the job.

One variant on the selective tendering system is that contractors are asked to provide information as to their status and capabilities, after which a short list of suitable contractors is drawn up for selective tendering. This system is frequently used by international clients.

Instead of asking for tenders the client may negotiate with one or more contractors to try to obtain a deal which will provide value for money. Negotiation is particularly appropriate when the documentation necessary for tendering is not yet ready, or where the client wants the contractors advice on aspects of the project. The final agreement may be based on cost-plus, a fixed or variable fee, a target price and shared benefits below this figure or on any variant of these.

These basic methods of selection may apply whatever the roles of the parties to the process. In the case of BOT the project could be put out to selective tender or the client might negotiate with a number of contractors. In the case of design and build there are problems in going out to tender because at the beginning of the process the exact nature of the project is unknown. If tenders are used the process is expensive because the outline design proposal has to be submitted at the tender stage. The contractor may alternatively be chosen by negotiation or because of some previous working relationship or recommendation. However, as the roles of the principal actor in the process embrace more and more functions, the cost of tendering escalates. In the case of PFI, for example, costs are very high and there has been discussion of the possibility of refunding some of the tendering costs of the short-listed service providers. There is, however, at the present time, no discussion of dropping or diluting the competitive element in the system.

In many countries of the Far East, the client and the contractor are often the same, because much housing and commercial building is undertaken by contractor developers on a speculative basis.

Partnering arrangements represent an attempt to remove confrontation from the relationship of the client and the various members of the building team. A continuing, knowledgeable client will thoroughly vet the professional practices and contractors hoping to participate in the partnership arrangements. In many cases, the potential team members will have worked with the client previously. The client selects those organisations he wishes to work with on a long-term basis. The process of selection may then be dispensed with, but there may still be a tender selection amongst those on the select list which would set the price for the project. This process is strongly recommended by the Egan report, but it is too early to say how well such arrangements will wear in practice. The costs of the initial vetting of the partners for the team are high, but they are low for each individual project. To put it in different terms, the initial transaction costs (see Chapter 10) are high and are a sort of fixed cost of this method of working, but the subsequent transaction costs, which vary with each additional project, are low so that if the client has a series of projects, the total transaction costs will be reduced.

Negotiation and partnership contracts accounted for about 12 per cent of all projects reported by quantity surveyors (DL&E, 1999). However, surprisingly, a survey undertaken of 85 large UK contractors found that two thirds of the top contractors obtained over 50 per cent of their work through negotiation (Shash, 1993).

The way the selection process works in practice varies according to the state of the market which in turn affects the relative power of the client and contractor. During the UK recession of the early 1990s when most contractors were very short of work, the client was able to drive a hard bargain, often, it is said, by negotiating downwards the price of the successful tenderer. In some cases the client made it known in advance what price he was willing to pay and effectively asked for bids below that figure. It is probable that a period in which the client is continually pushing for improvement in the service he receives and in the price he pays has a long-term beneficial effect on the efficiency of the industry. It is less certain what the effects are likely to be if the buyers market continues over a long period. Many large clients of the industry have an interest in obtaining the best bargain for their project. They are not interested in the long-term health of the industry. Continuing clients should be so interested, but the pressures imposed by financial markets favour profits now, rather than possible continuing profits in the longer term. The short term private interests of the client do not necessarily coincide with the long term interests of society.

The above procurement methods refer to the selection of the main contractor. In the UK, and to a lesser extent in other countries, the majority of work on large buildings is subcontracted so that it is the subcontractors, not the main contractor, who actually carry out the work on site. Both the role and the method of selection of the subcontractor are varied. While, in general, the subcontractor does not undertake design, for certain types of work he may be commissioned to provide detailed drawings for the components he is to supply. This is very usual for the installation of complex services for which only the subcontractor has the necessary expertise. In some projects a subcontractor may be nominated by the client or his design team. Such a subcontractor has a better bargaining position on price and the contract conditions than a subcontractor selected by competition. The selection of other subcontractors may be by selective tender or by negotiation. As in the case of the main contractor, the state of the market has a significant effect on the way the price is settled and on the contract conditions. Negotiation after the tender and pre-announcement of a maximum price are not uncommon. A major bone of contention between main and subcontractors is the 'pay when paid' clause under which the subcontractor does not receive his payment on the monthly certificate until after the contractor is paid, so that in some cases the subcontractor is supplying some of the working capital of the main contractor. Unfortunately, the provisions of the Housing Grants, Construction and Regeneration Act 1996 have made little practical difference to this situation.

In Japan nearly all the work is undertaken by trade contractors who maintain a special relationship with a main contractor who endeavours to provide continuous employment for his subcontractors. Price is determined by the main contractor, who monitors financial and project performance (DLSI, 1997).

The arrangements for terms of payment of the contractor will be very different according to the roles of the parties. Even for fairly traditional methods there are various options. In general, contracts are to undertake work for a fixed price or a predetermined method of fixing the price. However, in some countries for certain contracts, especially those of long duration, the contractor may be recompensed for changes in the price of materials and labour, that is, a fluctuations clause is included in the contract. Such an arrangement is possible only if there are accepted means of verifying prices.

An alternative contractual arrangement is that a schedule of rates is agreed for each item of work. The project is measured after the work has been undertaken and the contractor paid according to the rates agreed. This arrangement is particularly useful when the precise work is not pre-determined.

As already mentioned, it is possible to agree a fee, often linked to an incentive to the contractor to perform well. Sometimes, the benefit for a satisfactory project is shared between the client and the contractor, giving an incentive to the client also to facilitate the construction process. Clearly, the method of remuneration for the contracting party for provision of any service is very varied. Chapter 14 deals in detail with price determination.

Part Three
The Supply of Construction

Introduction

The supply of construction is provided by consultancy practices, firms from many industries including contractors and material and plant manufacturers, and individuals. Hillebrandt (1984) attempted to describe the broader industry as it was functioning in the UK in the early 1980s. In this book the emphasis is placed on contracting organisations of various types. They, with the labour they employ, account for about 40 per cent of the output of the industry (ONS, 1998, Table 2.2), so that while they are the single most important element, the other participants in the process must not be forgotten.

After a discussion of the objectives of the firm in Chapter 9, in Chapter 10 the costs of the construction firm are examined in some detail with the objective of understanding the supply curve of the individual firm. This is followed in Chapter 11 by seeing the implications of adding up the firms' supply curves to arrive at market supply curves; that is, what output the market is willing to supply at various prices. Chapter 12 links the cost and revenue curves of the market and the firm in various market situations and provides the theory necessary to take the analysis on in Chapter 13 to describe the demand curves actually facing the individual firm. Chapter 14, against the background of the realities of the cost curves of the firm and the market conditions in which it is operating, considers the way the firm decides how to respond to an opportunity to obtain work and how it fixes its price. Lastly, in Chapter 15, the analysis of costs and revenue is drawn together to discuss the equilibrium position of the contracting firm.

It will be noted that there is a major difference in the treatment in the theory of demand and supply. In the case of demand, the overall demand is studied and the demand facing the firm is deduced from the total situation. In the case of supply, the supply curve of the firm is analysed first and the total supply position is obtained by considering the aggregation of the situations of the individual firms in the market.

9
Objectives of the Firm

Economists use the word 'firm' to mean a business unit. The person who takes the decisions and risks of business is known as the 'entrepreneur'.

There are many types of firm ranging from a one-man business, through a partnership to a limited liability company and a public limited company in which the public may buy shares. These are the terms adopted in the UK but other countries have a similar range. The partnership is the form of firm used mostly by professional practices, although some of the largest professional firms are now limited liability companies. Contractors may be any of the types listed. In the UK in 1997 there were about 160,000 private contracting firms (DETR, 1998b) of which eight (nine in 1998) had a total turnover of more than £1,000 million. Five of these (seven in 1998) had a turnover in contracting alone of more than £1,000 million. About thirty further firms had a contracting turnover greater than £100 million (Building, 1998; Building , 1999). At the other extreme are the very small firms, often sole traders, which comprise most of those in the industry. Most of the labour-only subcontractors are in this category. Market economies, of whatever level of development, generally have a large number of small firms and a few relatively large ones. Australia has about 100,000 firms, of which only about twenty operate all over the country. Average employment is four persons (DLSI, 1997). Japan in 1997 had about 56,500 licensed contractors, including five very large ones (DL&E, 2000). The USA has over a million firms with an average employment of 3.5 persons. In addition, there are another 15 million working partners or self-employed proprietors (DL&E, 2000). Turkey has about 55,000 registered contractors (DL&E, 2000). Planned economies, including the former Soviet Union have, or had, almost entirely large firms. In Vietnam in 1993 there were only 360 construction units under central government and in 1995 it was estimated that there might be up to 1,000 small-scale private firms (DLSI, 1995). Whereas in the former Soviet Union (FSU) there were only about 2,000 construction organisations, by 1993 there were 60,000 construction organisations in Russia, including building material producers (Russian Construction Research Group, 1993). In Latvia, formerly part of the FSU, by 1994 there were approximately 800 companies operating as contractors (Segal Quince Wicksteed, 1995).

The vision of the business

The 'objective of the firm' is usually taken to mean a relatively quantifiable goal often linked to a certain period of time. There is, however, a vaguer concept of what the firm wants to achieve which is often referred to as the vision or mission of the business. Ramsay (1989) defines mission of a business as 'its long term ambition, what it *ideally* wants to become over time. This is usually expressed in qualitative terms and is probably never achieved exactly or totally.' Mission can take many different forms. It may, for example, be to become an international contractor; to move from being a general builder to being a civil engineering contractor; to have a first class reputation within the industry and the community; to enhance the business to sell and be able to retire, or to keep the business small enough and profitable enough to be run at a senior level by two or three family members. Most firms have a vision or a sort of dream of the future, but it is not always acknowledged even by the main players. It is helpful to bring it into the open because it can then serve as a litmus paper by which to set and judge objectives.

This book is mainly about the contracting firm and industry but many contracting firms are in fact in other businesses, sometimes as a result of historical accident, sometimes as a result of deliberate policy. The businesses that a firm is in are often closely linked to its vision of the business. The decision of what business to be in is examined from a theoretical point of view in Chapter 17.

Objectives of the firm

With so many possible visions of a business, it is not surprising that the range of possible objectives is similarly great. Archibald (1987), in a comprehensive review of models of the objectives of the firm, classifies models as optimising and others. The optimising models include the traditional microeconomic objective of the entrepreneur of the firm in neoclassical economics as the maximisation of money profits. It still forms the basis of much theory of the firm and, together with the marginal approach which it embodies, and using a process of aggregation of business units, it enables statements to be made about total market phenomena and equilibrium conditions.

The profit maximisation assumption has been challenged at many levels. The first is that it does not take account of the other objectives of the firm. This is a very valid objection. However, it is also true that profit, even if not absolute maximisation of profit, is a requirement for a firm to stay in business and that it is important to understand the theoretical basis of profit maximisation, even if it is not going to be applied in its entirety. Baumol (1959) maintains that firms aim at sales maximisation complemented by a satisfactory level of profit. This is especially important when the firm is run by managers, rather than the owners of the firm. Simon (1959) too stresses the likely difference in objectives of a firm run by its owners and one run by managers, with the latter much less concerned with profit maximisation.

At the more trivial level, for example, it is pointed out that there is no reason to suppose that the entrepreneur will, above a certain level of profit, value increases in profit more than increases in his leisure or that his objectives may change from making money to gaining respect and, say, becoming a local dignitary or secretary of the golf club. It is not difficult to accept this comment and still maintain that the overall concept of profit maximisation as the objective is tenable – allowing some idiosyncrasies for the small firm. Even the problem that if a high degree of competition exists, any entrepreneur who does not maximise profit would go out of business, can be overcome by postulating that the level of income which he is prepared to accept is lower than that which he could earn elsewhere. There is also considerable support in the contracting industry for the idea that contracting is a 'way of life' and that many contractors would not wish to cease business even if they could obtain a higher return on their capital and labour by using it in some other way. This is, presumably, true in one degree or another in a great many industries, for example, certainly in agriculture, but the distinct characteristics of construction lead one to suppose that it may be especially true of this industry. It may also be true that many family firms may limit their expansion to maintain the firm at a size manageable within the family. Thus, it could be that sometimes the expansion of a family firm is determined by the number of children growing up and wishing to work in it.

The second objection to profit maximisation is that it bears little relation to what firms actually do, because they do not have enough knowledge. Simon (1959), in a paper which is still influencing the debate on this issue, points out that entrepreneurs have limited knowledge and computational ability and are therefore subject to 'bounded rationality'.

A third group of objections to the supposed drive for profit maximisation was led by behavioural economists who question the assumptions of neoclassical economics and, observing the behaviour of firms, found that they simply do not maximise. Cyert and March (1963) and, to some extent, Baumol (1959) and Williamson (1975) belong to this school of thought. Cyert and March see the objectives of the firm, mainly the large corporations, as determined by organisational structure and the internal operations of the firm, as much as by monetary objectives. Williamson stresses the influence of transaction costs in influencing the way firms behave.

Hillebrandt and Cannon (1990) found in their study of the strategy of large construction firms that they had to discuss the vision of the business and objectives together and 'divided them into three broad categories: (i) financial; (ii) quality of performance or relationships, and (iii) size and type of business'. It was found that in companies which are part of a conglomerate, the maximisation of profit or cash flow was dominant. The family firms had a wide range of objectives and the word 'maximise' was hardly used in the interviews. The remaining firms also had a wide range of objectives, but with rather more emphasis on finance and profitable growth than in the family firms. Overall, they found that profit and other financial objectives certainly played a substantial part in the thinking of firms. It would seem that, firstly, the greater the convergence of interests of owner and manager, the greater the likelihood of some approximation to the objective of profit maximisation, and, secondly, that the greater the degree of competition, the more important it is for survival that entrepreneurs aim at profit maximisation.

In the chapters that follow the objective of profit maximisation will be considered first. A study of the theory which relates to this objective has the added advantage that there is a whole body of microeconomic analysis, including an analysis of market situations, which is founded on marginal analysis and the objective of profit maximisation. It would be foolish to reject all this analysis without understanding the messages it has to give.

The measure of profit in which entrepreneurs in most industries are interested is return on capital employed, firstly because it is this which the owners of the firm require, and secondly because it is this profit which should be compared to returns on capital in alternative uses, such as other industries or fixed interest securities. In the analysis which follows however, profit is often considered in relation to turnover. Maximum profit on turnover will produce maximum profit on capital only if the ratio of capital to turnover is the same in all parts of the business. In fact, there is a substantial variation in the amount of capital required for different construction-related activities. Contractors generally think in terms of profit in relation to the value of a project, and hence of profit in relation to turnover.

There is a further problem, namely that, in contracting, capital for a project or group of projects may be nil or negative. This is possible, mainly for large projects, in the UK because:

- the only fixed capital employed on site is plant and equipment and this can be hired rather than owned;
- contractors are paid on monthly certificates on the amount of work done. If they are paid on time, this reduces the working capital required;
- money outstanding is further reduced by extended credit given by builders' merchants, which may mean that money for materials does not have to be paid out until after the contractor has been paid for them;
- main contractors often do not pay subcontractors for work done until after they have been paid for that work. As a large part of the work of the industry is subcontracted, this reduces outgoings of main contractors very substantially. It works, of course, the other way for subcontractors;
- there may be, especially on large projects, an initial payment to contractors before work commences;

• amongst other favourable pricing arrangements, it may be possible to get paid more for work done at the beginning of a contract
by pricing the work which is done early in the construction process
higher than work done later, so that the income is high at the
beginning of the contract. This is known as 'front end loading'.
This particular practice should, however, be minimised by com-
petent supervision by consultants.

If the capital required is nil or negative, the return on capital
employed is infinite. This concept is difficult to handle in theoretical
terms.

To set against these reasons for low working capital is the retention of around 5 per cent, with a retention period of say twelve
months. This substantially increases the working capital required
and may well eliminate any positive cash flow. In addition, especially in civil engineering, claims take a long time to settle so that
cash is reduced. In countries other than the UK, it is unusual to
have very low working capital requirements. Indeed, in developing
countries a shortage of working capital is one of the principal constraints on the development of contractors.

Return on capital employed is important when the choice of
businesses is being considered and in this analysis too availability
of other inputs, notably management, is also a vital factor (see
Chapter 17).

The second objective to be considered in this book is that of
making a normal profit or some other minimum level of profit and
thereafter seeking maximisation of turnover. This assumption seems
to meet some of the problems of different objectives of owners and
managers. It then becomes possible to compare the effects of the
two different objectives.

A high positive cash flow or minimisation of negative cash flow
must also be considered as a secondary objective. A positive cash
flow is an important means to making a profit. At times in the UK
it has become a dominant objective, especially in periods of shortage of work. There have been times when contractors have made
more profit by investing their positive cash flow than from mark-
ups on the contracts. Moreover, many insolvencies of construction
companies are triggered by a lack of cash, rather than long-term
negative profits.

10
Costs of the Construction Firm

Types of cost relationship

One of the peculiarities of the construction industry is that work is obtained in the form of contracts for projects which are large and indivisible but that the work load relating to each project is spread over a long period of time. Costs (and revenue: see Chapter 13) have therefore to be examined in three distinct ways. Firstly, the cost of the project as a whole; secondly, the cost of the total project must be related to the work load over time; and lastly, the cost of various alternative work loads at a given point in time must be analysed. The usual cost curves of economic analysis are of this last type. The method of transition from the first to the third will be discussed first.

Figure 10.1 shows the estimated total cost of the contracts received at various dates and the period of time over which the work will be carried out. Usually, the work on a contract builds up over time first slowly and then more rapidly to a peak and then declines, first quickly and then slowly, prior to final completion. However, for simplicity of exposition it is assumed here that work is spread evenly over the period of the contract starting in the month following the order. Thus, contract (1) for £12 million obtained in January and lasting 60 months costs £200,000 a month from February to December in the same year. Contract (2) for £2 million in April spread over 20 months costs £100,000 a month from May to December, and together contracts (1) and (2) have a cost shown by the line *ABCK* in Figure 10.2. Two other contracts obtained in the year lead to a cumulative cost curve for all four contracts of *ABCDEFGH*. At the same time there will be work going

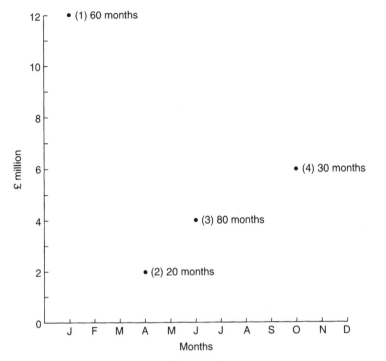

Figure 10.1 Contracts obtained during the year

on from contracts obtained in previous years and this is represented by the stepped dotted line from *L* to *M*. Altogether the cost of work over the year is shown by the top line from *L* to *H*. Thus, the extremely erratic discontinuous new contracts situation yields a less erratic continuous cost curve over time, although still with substantial ups and downs.

If a month is taken as a sufficiently short period of time to be regarded as a 'point in time' over which the usual cost curves of the firm in economic analysis are relevant, in the month of June the firm would be at a point on its cost curve shown by the point *N* in Figure 10.2, with an output of £500,000 units of work and a cost of £500,000. For simplicity, let the £500,000 of units of work be regarded as units of output at a unit cost of £100,000. The question to which the firm requires an answer is: how would the cost be affected if, instead of producing 5 units in June, it was producing a larger or smaller amount? This is a very relevant question to the contracting firm because the output in June is partly determined

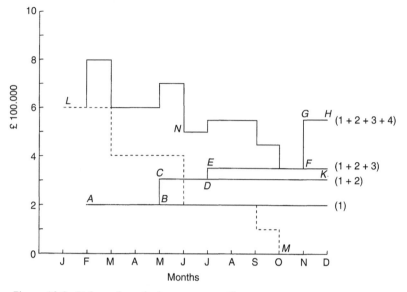

Figure 10.2 Value of work done per month

far in advance and therefore is something which the firm can hope to control and yet, for reasons described later in this chapter under the heading of 'Uncertainty in Work Load', it may have a higher or lower output thrust upon it by circumstances outside its control.

The answer to the question of the shape of the firm's cost curve will be different according to whether the period under consideration is short, so that the overhead expenses and hence facilities of the firm cannot be altered, or whether the period is so long that the whole structure of the firm and the level of technology at which it operates can be adjusted. Moreover, the answer will be different according to the work mix of the changed work load. It could be that a higher or lower output is achieved either by changing the average size of contract and keeping the number of contracts constant or by changing the number of contracts and keeping average size of contract constant. In practice, of course, there will probably be a combination of the two, but for analytical purposes it is useful to consider the two extremes. In the discussion of inputs of site construction and the contracting firm which follow, attention is concentrated first on the relatively short-term position in which the expenses at head office are fixed and the supply of certain site management is fixed. Inputs are divided into three categories: variable, which vary directly with the value of output and which must be covered in the short run; fixed, whose level cannot be altered in the period under review but which must be covered if the enterprise as a whole is not to make a loss; and postponable costs, which must be covered in the very long run but need not be covered in the short run. It will be assumed in the analysis that, apart from variations in size and number of contracts, the type of work done by the firm can be expressed in homogeneous units. It is also assumed that the level of output from which divergence is being considered is within a band of optimum or near-optimum combinations of resources for their level of output. Table 10.1 summarises the broad conclusions derived from the discussion below under each input of the changes in the level of output. Costs are expressed in terms of cost per unit of output. If total cost changes in proportion to output, then the cost per unit of output will be the same. If total cost rises by more than the output, costs per unit of output will rise, and if total cost rises less than proportionately to output, then costs per unit of output will fall. These relationships are discussed more fully under the heading of 'Short-run Total Cost Curve and the Derivation of Average and Marginal Cost Curves'.

The costs referred to above are costs of production. They will be discussed further in this chapter under the headings of variable costs and fixed costs. First, another type of cost, transaction cost, will be described.

Transaction costs

Attention was drawn to the costs of making transactions in a seminal article by Ronald Coase (1937) and their implications further explored by Williamson (1975). Both were explaining why firms exist, because it might be thought a simpler arrangement for the production process, for each specialist to sell his piece of work on to the person engaged in the next activity. The reason is that transactions are too complicated and expensive. Gruneberg and Ive (2000) list the following transaction costs which affect construction, although they do not claim that the list is exhaustive:

- Search costs are the costs of finding out information about who is offering what products or services and at what price.
- Product or service specification costs arise because the thing being purchased may not exist when it is bought, so that precise definition of what is being bought is essential.

100

Table 10.1 Change in short-run unit costs below and above output at or near optimum level of output

	Change in output by change in size of contracts		Change in output by change in number of contracts	
	Below optimum	Above optimum	Below optimum	Above optimum
Variable costs				
1. Materials used on site	Higher	Same then higher with management inefficiency	Same or higher if no bulk buying	Same then higher with management inefficiency
2. Labour used on site	Higher	Higher with over-time and management inefficiency	Same	Same then higher with management inefficiency
3. Some site management	Same	Same	Same	Same
4. Plant and equipment on site	Same	Same	Same	Same
5. Interest on working capital for site work	Same	Same	Same	Same
6. Estimating costs assuming constant ratio of successful bids	Same	Same	Same	Same
Fixed costs				
7. Employees at head office including contracts and some site managers	Higher	Lower	Higher	Lower
8. Outgoings on buildings and equipment	Higher	Lower	Higher	Lower
9. Minimum level of remuneration of entrepreneur	Higher	Lower	Higher	Lower
10. Interest on loans if not postponable	Higher	Lower	Higher	Lower
Postponable costs				
11. Normal remuneration over minimum of entrepreneur	Higher	Lower	Higher	Lower
12. Normal return on capital which can be withdrawn	Higher	Lower	Higher	Lower

- Contract selection, contract design and negotiation costs again arise because the product of the construction industry does not exist when it is bought and because of the great diversity of product.
- Supplier selection costs occur because there is no single uniform market price for the relevant product or service, for example, for the choice of subcontractors.
- Contract performance monitoring costs are transaction costs which are incurred due to the need to measure and control performance in terms of its price, timing and quality.
- Contract enforcement costs are the costs of legal bills and delays.

It may be that the increased use of electronic communication will reduce transaction costs, for example, in ordering materials, but many of the costs are not readily susceptible to these methods.

Transaction costs arise in construction particularly because of the complexity and diversity of the process which means that certain specialisms are required rarely, so that it is not worthwhile any but the very largest firms having them in-house. The existence of professional bodies with a distinct view of the roles of their members in relation to the construction process and their wish for them to be independent of commercial organisations has historically hampered integration of construction processes and led to the continued separation of design and construction. These arrangements are eroding somewhat at the very top of the industry but are likely to remain for the great mass of operations.

Variable costs

Materials used on site

Materials and components used on site account for around half the total cost of construction. In broad terms the cost of materials will vary directly with the amount of work under way. There might be some exceptions to this at extremely low sizes of contract because of the lower limit to the quantity in which certain materials can be bought, the non-availability for small purchases of quantity discounts and the possibility that central purchasing may not be worthwhile. If the number of contracts is small, the size remaining the same, the effect of small outputs on costs of materials is likely to be negligible unless the low turnover implies the abandonment of central bulk buying.

Assuming that at the original output there was already centralised buying and that quantity discounts were obtained, then higher levels of output either with more contracts or larger contracts should not affect the unit cost of materials until the point is reached where the management of the project becomes inefficient owing to fixity of some management skills leading to wastage and damage of materials and components.

Labour used on site

Site labour accounts for about a third of the cost of construction. It is employed by main contractors, subcontractors who supply and fix the materials of their trade, and labour-only subcontractors who supply only their labour. The last arrangement is common in the UK and a number of other countries, for example Singapore (Ofori and Debrah, 1998), but in some countries it is discouraged or even illegal.

Where labour-only subcontracting is used the operatives are self-employed and must be found, engaged to undertake the work, often on a piece rate basis, supervised and checked. Transaction costs might be expected to make this sort of arrangement prohibitively expensive. Buckley and Enderwick (1989) examined transaction cost theory in relation to the labour markets in the construction industry and came to the conclusion that theory would suggest that an internal labour market (that is, directly employed labour) would be the most efficient. In the UK, however, there has been a steady increase in self-employment in the industry, showing that labour-only subcontracting has been preferred and contractors believe it to be cheaper than direct employment. There are several reasons for this. The first is that employers' dues to government in the form of national insurance contributions and so on have often been lower, although in the UK this situation is changing. Operatives are prepared to accept and enjoy the more favourable tax position and greater independence. The most important reason is, however, probably the greater flexibility it brings in the organisation of the operative labour force in the face of large lumps of discontinuous work which fluctuates over time. This conclusion is supported by Buckley and Enderwick (1989) and also by the related fieldwork (Hillebrandt and Cannon, 1990). The alternative is almost certainly a work force where more persons are underemployed some of the time.

In Japan, where main contractors have a special relationship with their subcontractors, they stipulate the contract price and monitor the subcontractors' financial and project performance (DLSI, 1997). This reduces transaction costs and to some extent gives many of the benefits of both direct employment and of subcontracting.

The costs of labour-only subcontracting vary with the state of the construction market. Whereas the wage rates for operatives directly employed by contractors are fixed in the short term, the rates paid to self-employed fluctuate directly with the demand and supply situation. Thus, if a contractor increases his output in line with a rising construction market, his costs per unit are likely to rise if he is using self-employed labour. There is little evidence as to what will happen to costs as output increases with directly employed labour, unless the increased output creates a deterioration in the quality of management causing inefficiency in the organisation of labour, in which case costs would rise. It would also rise if there were a shortage of labour in the organisation leading to expensive overtime working.

Site management

Traditionally in the UK senior management, including some site management, remains with a company on a long-term basis with mutual loyalties between the employer and employee. This tradition is breaking down and some site managers are now employed on contract for the duration of the construction contract on which they are working. In other countries there is a wide range of practice, with Japan at one extreme having had an assumption of a job for life and other countries, including some developing countries, having site managers with a high degree of job mobility.

The costs of site management will depend partly, as with other inputs, on whether there is an overall shortage of good managers. If there is not, there seems no particular reason why it should be expected that the cost of variable site management would not rise roughly in proportion to the work done. In the case of large contracts, personnel in charge would be paid a higher salary, and in the case of a larger number of contracts management personnel would be duplicated. In any case, the absolute cost of management is relatively small in relation to the total cost of the contract.

The effect and importance of management lies not in the direct costs but in the effect it has on the efficiency of the whole project and hence on labour costs, and even on material and plant costs. This is particularly relevant as management expertise is one of the scarcest resources of the construction industry in the UK, in many other developed countries and certainly in developing countries.

Plant and equipment on site

Plant and equipment used on site may be hired or owned. If it is hired its cost is the hire fee. If it is owned, its cost has two components: that which is dependent on the extent to which it is used, that is, depreciation which varies with intensity of production, and that which depends on the rate at which it becomes obsolete. Its capital cost should be recouped over the total contracts for which the equipment is used. If, as is the case on many large contracts, the item of plant and equipment is worn out during that contract, then the whole of the cost has to be related to that particular contract. If it is used over several contracts, then the cost not due to its use is more properly considered as a fixed cost.

The amount of plant used will increase with size of contract, but there is little evidence to indicate whether or not it will increase proportionately. An assumption of a change in plant costs proportionate to output is not unrealistic for output changes with constant and changing sizes of contract.

The situation with plant and equipment on site may change dramatically in the future, however. In Japan, some large contractors have developed integrated site construction systems which require the minimum of site workers and these systems have been used outside Japan, for example, in Singapore. In such cases, plant and equipment will constitute a huge cost, part of which will probably be specific to their project and part to a number of projects.

Working capital for site work

Working capital or any other finance is not a resource of construction in the same way as, say, materials or labour. Indeed, if the money invested were added to the other inputs this would result in double counting, because if it is invested it is always in some physical resource which would have been considered in any case. Nevertheless, it is necessary for the construction firm to be able to use resources before they are paid for, and the interest paid on working capital is the cost of this facility.

The normal method of payment for construction contracts is that once a month the architect or other professional adviser to the client certifies the value of the work which has been done on site and the client pays the contractor the amount of the certificate minus a small percentage which is retained until after the end of the contract in case of defects in the work. The contractor needs working capital to pay for material, labour, etc., from the time of commencement of work or from the date of the last certificate until the date of the next certificate, plus the period between the date of the certificate and his receipt of payment. There is room for considerable skill in keeping this working capital to a very low figure by, for example, the use of credit from builders' merchants and other methods outside the scope of this book. There may even be instances where the main contractor manages to ensure that his cash receipts exceed his cash outgoings, in which case working capital required is negative. This is common in the UK and in this case the main contractor is able to earn interest on his positive cash flow. The subcontractors in the UK are not in such a favourable position. Their payment is very often delayed and they, being dependent on main contractors for work, are not in a strong position to insist on the letter of the contract. However, there is a positive requirement for working capital in most countries, and its cost will be the rate of interest the contractor has to pay on it (or the forgone interest if he is using his own capital). The actual sum involved is negligible in most contracts, compared with the cost of labour and materials.

The importance of working capital in the industry is, however, very great, not on account of its cost but because contractors may be unable to obtain sufficient working capital at any price. This is mentioned later, in a consideration of the rate at which construction firms can grow.

It is noteworthy that some procurement practices require considerably more capital than traditional processes. This is especially so in countries where the contractor is to a large extent the developer both for housing and for commercial projects, as, for example, in Hong Kong, Malaysia, Singapore, Thailand and Indonesia. In this case speculative building is the norm.

Costs of estimating

Assuming a constant success rate, estimating costs are probably more or less proportionate to the value of the work obtained, whether in large or small contracts. For this reason they are included under variable costs, although estimators are normally part of the head office establishment. After the business has been obtained, however, they are fixed costs because, as was mentioned earlier, the cost of estimating for successful and unsuccessful tenders is borne by the successful tenders. The higher the success rate in bidding, the lower the costs of estimating which have to be added to each contract. According to Flanagan and Norman (1989), discussions with contractors indicate that they expend in the order of 0.7–1.0 per cent of turnover in the handling of tender documentation. As non-traditional procurement systems such as PFI increase, this figure has risen substantially, and is likely to rise further.

Fixed costs

Fixed costs are a relatively small proportion of total costs of a construction firm and consist of the expenses incurred at head office, plus obsolescence and depreciation not related directly to use of plant whose life extends over more than one project, and site management which is difficult to recruit easily and for whom the business offers security of employment. They include too a minimum level of remuneration for the entrepreneur as well as interest on loans (other than those already considered for site working capital) the payment of which cannot be postponed.

Expenses at head office have changed over the recent past, with an increasing cost of management systems for quality and environmental management and computer systems. They include research and development costs, which are generally low, but which, with the increase in non-traditional procurement arrangements, may well increase. In Japan large contractors spend about 1 per cent of turnover in-house on research and development, and they also commission outside research (DLSI, 1997). With the newer procurement arrangements an increasing variety and depth of expertise is required to market and manage projects, especially where the contractor's role extends backwards to the provision of finance and design and forwards to the operation of the asset.

Since these costs are all by definition regarded as fixed, and therefore do not vary with the amount of work undertaken, they will be lower per unit of output the higher the level of output and vice versa, irrespective of whether the output is in large, medium or small contracts.

The remuneration of the owners of the firm who are working in the firm is often considered as a part of profit by accountants. Economists consider normal profit as that profit which is just enough to keep the entrepreneur in the industry, as a long-run cost of the business. This normal profit would be closely related to that which he could earn elsewhere, although not necessarily synonymous with it for reasons discussed in Chapter 9. However, it will normally be larger for the more efficient producers than for the less efficient ones. In the short run, however, the entrepreneur would usually be prepared to stay in the industry for considerably less than normal profit. If he had no other earnings he would, however, presumably need some minimum level of remuneration, and it is this which is included in fixed costs.

In the case of fixed investment in plant and buildings, the problem of the difference between the physical assets and the financial position arises again. Fixed investment in the buildings and equipment which is already owned by the firm is, in the short run at least, a sunk cost (see Chapter 2), which need not be considered. However, the extent to which this is so in practice depends on how the physical assets are financed. If they are financed by loans on which interest payments cannot be postponed, then the interest charges must be included and this will be a fixed cost, since it has already been assumed that in the short run it cannot be altered. If, on the other hand, it is, for example, financed out of equity capital on which there is no obligation to pay any return, then the cost can be postponed. Similar arguments apply to the working capital required.

Postponable costs

Costs which may be postponed are that part of the normal remuneration of the entrepreneur which is not required by him to maintain him and his family (which is a fixed cost) and the normal return on all capital invested in the business, apart from that on which interest has to be regularly paid. If the entrepreneur is content to allow the business to remain at its present size or is incapable of expanding it, then only the normal return on capital which can be withdrawn from the business must be met. The remainder is a sunk cost which does not affect the operation of the business at its present level. In the long run, if the business is to expand the whole of both these must be met because otherwise it will not be possible to obtain further finance for expansion. If no return is accruing to existing equity capital, the public will not be prepared to provide any more of it.

The entrepreneur is unlikely to regard the profit which he should earn to keep himself in the industry as directly proportionate to output, but it seems likely that when the business expands to a higher level of work and employs more expensive specialists he will feel that he is now in charge of a business of a different type and will set his sights higher so far as his own remuneration is concerned. Thus, it is likely to go up in step fashion. In the short run, however, he will accept a fixed level.

The normal return on capital in the long run is also likely to go up in step fashion as the investment is unlikely to expand smoothly, e.g. expenditure on a new office block comes as a large once-and-for-all expenditure. However, in the short run it has been assumed to be fixed.

Short-run total cost curve and the derivation of average and marginal cost curves

From the discussion above, it is apparent that the total short-run variable cost of producing low levels of output is fairly high and increases rapidly at first, then less quickly at medium levels of output. It may then increase more than proportionately to increases in output because of the fixity in the short run of the supply of certain types of management and the facilities provided by head office. Table 10.2 column (3) shows how variable total costs might change with the level of output, and this is shown in graphical form in Figure 10.3A by curve *TCV*. The values of the changes shown probably exaggerate the extent of the variability in the normal firm but have been shown in this way for ease of exposition.

Column (2) of Table 10.2 shows the short-term fixed costs as being £100,000 and in Figure 10.3A these are the straight line *TCF*. Column (4) in Table 10.2 and curve *TCT* in Figure 10.3A show the sum of the total costs.

Average variable costs are obtained by dividing column (3) by column (1) in the table and are shown in column (6). They decrease to a minimum at 4 and 5 units of output and then increase again. Average fixed costs are similarly derived from columns (2) and (1) and are shown in column (5), and total average costs may either be derived from the total of total costs or from the sum of fixed and variable average costs.

Table 10.2 Short-run total average and marginal costs of the firm (£000)

| (1) | (2) | (3) Total costs | | (4) | (5) | (6) Average costs | | (7) | (8) | (9) Marginal costs |
Output	Fixed	Variable	Total		Fixed	Variable	Total		Extra cost of 1 unit	Smoothed marginal
0	100	0	100	—	—	—	—	—	—	—
1	100	90	190	100	90	190	90	90	85
2	100	170	270	50	85	135	80	80	78
3	100	246	346	33	82	115	76	76	75
4	100	320	420	25	80	105	74	74	77
5	100	400	500	20	80	100	80	80	101
6	100	522	622	17	87	104	122	122	122–5
7	100	665	765	14	95	109	143	143	183
8	100	888	988	12–5	111	123–5	223	223	266
9	100	1,197	1,297	11	133	144	309	309	381
10	100	1,640	1,740	10	164	174	443	443	376

Average costs may be derived geometrically from the total cost curves and are represented by the slope of the cord in Figure 10.3A. Thus the average total cost at 3 units of output is shown in the slope of *OA*, which equals the total cost of 346 divided by 3 units of output = AB/OB. Average fixed costs are shown by the slope of *OC* and *OD*, etc., and clearly decrease as the units of output increase. The various types of average cost are plotted in Figure 10.3B: *ACV* denoting average variable cost first decreasing and then increasing, as does the slope of the cord of *TCV* in Figure 10.3A; *ACF*, a rectangular hyperbola, denoting average fixed costs decreasing along its length; and *ACT* being the sum of the two.

Marginal costs are the increase in total cost due to the last small increase in output. Since total fixed cost is constant, marginal fixed cost is nil and marginal cost is therefore entirely dependent on variable costs, that is, those relating to the project under the heading 'variable' in Table 10.1. The additional cost attributable to the last unit of output is shown in column (8) of Table 10.2. However, because the units of output as defined are so large, this is not sufficiently fine a measure for the analysis required, and marginal cost must be defined in terms of infinitesimally small increments in output, or, in geometric terms by the slope of the tangent to the total cost curve. This has been approximated in column (9) by taking the difference between the total cost at one unit below and one unit above the point required, divided by 2. This marginal cost is plotted in Figure 10.3B, and is at its lowest at output 3. It is also clear that the slope of the cord and the slope of the tangent to the total cost curve in Figure 10.3A are the same at output 5, that is, at *E*, and this is the point at which the cord has its lowest slope. On Figure 10.3B this is where the marginal cost curve and the average total cost curve cut. Thus, the marginal cost curve always cuts the average cost curve at the latter's lowest point.

It will be noted that at 5 units of output the average total cost is £100,000. These are the figures which relate to the output of the contracting firm in June derived from Figure 10.1 and 10.2. Thus, the question posed at this stage as to what the costs would be at outputs above and below this level has been answered by the general shape of the cost curves in Figures 10.3A–B.

Three general rules as to the relationship between marginal cost and average cost may be stated: (i) when the average cost curve is rising, the marginal cost must be below it; (ii) marginal cost equals

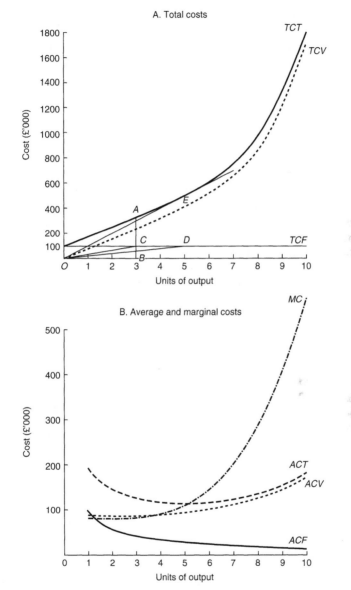

Figure 10.3 Short-run costs of the firm

average cost when average cost is at its lowest; and (iii) when average cost is rising, marginal cost must be above average cost. Thus total, average and marginal costs are ways of expressing the overall cost situation. The one which is chosen for discussion at any time will depend on the nature of the analysis being undertaken.

Difference between traditional and prefabricated construction

The general relationships assumed in these diagrams apply to the contracting firm undertaking traditional construction work. If, however, the firm is producing major components in factories off-site, it will have cost relationships more akin to manufacturing industry. Fixed costs will be a much higher proportion of total costs. This means that at small levels of output the fixed-cost element will be very high and the slope of the average cost curve before the lowest average cost will be much greater. Variable costs will be correspondingly less important. Thus, so long as the firm is not producing at full manufacturing production capacity, average variable costs are likely to decrease or increase only slowly even if there is some increase in variable costs due to the fixity of management, etc., in the short run. When the firm is approaching full manufacturing capacity, i.e. the maximum sustainable output per unit of time, it may be able in the short run to increase output by increasing overtime and by some neglect of repairs. For this increase in output, however, variable costs are likely to rise quite steeply. There will come a point when the plant simply cannot produce a larger output, and beyond that point the cost curves of the firm in the short run will be quite irrelevant. There is no such absolute cut-off limit for contracting firms because for quite large ranges of output management can be stretched, and in any case the variable inputs, being such a large proportion, can be increased to obtain a larger output. As has been indicated, however, the cost of such an extension of output may be considerable.

Uncertainty in work load

In a firm where most of the work is obtained on the basis of tenders, it is not possible to anticipate very far in advance exactly what proportion of the tenders put in will be successful. In each company there will be a generally assumed level of, say, 1 in 5 or

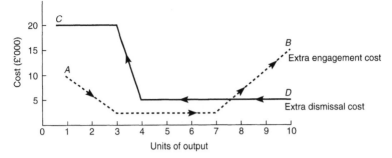

Figure 10.4 Non-reversible marginal cost curves due to extra costs of engagement or dismissal

1 in 10. The success rate can change quickly because of external changes in the market situation not anticipated by the firm, or by a small change in policy on the part of the firm itself. If this happens, its actual output may be different from that planned and the number of contracts, which is one of the major determinants of management needs, will rise (or fall) faster still. Thus, it is relatively easy for a contracting company to reach the level of rapidly increasing short-run average and marginal costs 'by mistake'. The danger of this arising will be greater, the greater the ratio of tenders put in to tenders accepted and the smaller the number of contracts normally undertaken at any one time. The distribution of types of selection amongst varying sizes of firm is not known but all sizes still have substantial reliance on tenders or uncertain selection processes. If, however, the contractor, instead of increasing his success rate, decreases it, he will be forced back along his cost curve and his average costs will rise. In this case the cost curve will rise faster with smaller outputs because the lower the success rate in bidding, the higher the costs of estimating which have to be added to each contract.

Non-reversibility of cost curves

The short-run cost curve is intended to show the effect on costs of *being at* various levels of output, assuming that the time period is too short to do anything about certain fixed items. In fact, however, the extra cost of increasing output by a given percentage and the reduction in cost due to decreasing output by the same percentage may not be identical. There are, for example, transaction costs of recruiting labour in advertisements, management time for selection, a period in which the man recruited is adjusting himself to the new job, etc. Some of these are virtually a form of fixed cost specific to recruitment, but they are marginal to the expansion process as a whole. They are plotted in Figure 10.4 at *AB*. They are high when few men are employed and rise again as more effort is required to obtain manpower as labour becomes scarce.

There are also costs in dismissing a man, and with the UK redundancy legislation if he had been in the employ of the firm for more than two years these costs might be considerable. It has been assumed in curve *CD* that the men dismissed first are not entitled to any redundancy pay, while some of those dismissed later are so entitled. These costs must be added to the original marginal cost curve to obtain a marginal cost curve for increases in output and then for decreases in output.

There will be similar cases of non-reversibility of curves with, for example, the purchase of new plant and the sale of secondhand plant or of materials from the manufacturer and of surplus materials.

Problem of accurate estimation of costs

In Chapter 14 the problem that the firm does not know its costs in advance is covered in more detail. It is, however, important here to realise that the level of the cost curve of the firm at any stage is subject to a wide margin of error. This can cause some wrong decisions, and indeed does do so, as is often seen by the wide difference between expected and actual profit. It does not, however, invalidate the analysis, which can be repeated with a 'band' of costs.

Long-run costs

There are two meanings of 'short run' as opposed to 'long run', namely the short run as the period in which certain costs are postponable and the short run as the period in which fixed plant and equipment and the head office organisation cannot be altered. The two categories overlap, since some postponable costs have only to be met in the long run if the level of fixed costs is to be altered. It is, however, helpful to consider their effects on the long-run situation separately.

Postponable costs have been isolated in Table 10.1 as the remaining part of the normal remuneration of the entrepreneur or, that necessary to keep him in the business and the normal return on capital which can be withdrawn. In the short run, capital was taken as given and it was assumed that the entrepreneur would not in a short period raise his required level of remuneration, and therefore these had the characteristics of fixed costs. In the slightly longer period these costs are not postponable, otherwise the entrepreneur and the capital will be withdrawn from the business. Moreover, in the still longer term the entrepreneur will wish to raise his level of remuneration more in keeping with the volume of business he is handling. Lastly, if fixed costs are increased, then the normal rate of return on the equity capital which cannot be withdrawn becomes relevant, for unless it is paid, further capital will not be forthcoming.

Thus, in the first use of the phrase 'long-term', that is, the sense in which costs are no longer postponable, the difference between the long-run cost curve and the short-run cost curve would be that fixed costs are higher. They are also likely to be stepped to take account of the increase required by the entrepreneur in the return on his labours.

In the second use of the expression 'long-term', fixed costs may be increased in combination with a change in technology or simply by, say, doubling the quantity of all inputs of the business. In practice, the latter is unlikely, since as a business grows in size it becomes more economical to change the proportion and types of inputs. This is discussed using iso-product curves in Chapter 16. Here, attention is focused on the effects of various levels of fixed costs.

Effect of different technologies with different fixed costs

In Figure 10.5 three levels of fixed investment are considered as represented by FC_1, FC_2 and FC_3. Each level of fixed cost is related to a variable cost curve (not illustrated) and hence to a total cost curve – TC_1, TC_2 and TC_3 respectively. It is apparent that in this highly simplified situation the solid curved line *ABCD* is that showing the lowest possible cost, and hence the different technologies and their related fixed costs should be introduced at 8.5 units of output and 13 units of output respectively.

It is interesting to note that at point *B* the same output can be produced at the same cost with two completely different levels of fixed costs and technologies. This explains why firms of two widely different sizes and organisations can compete effectively for the same contract.

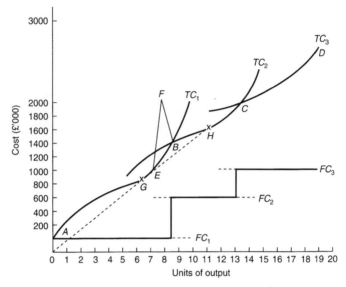

Figure 10.5   Long-run total cost curves with varying fixed costs

Figure 10.5 is in fact an over-simplification. Firstly, any investment in fixed plant and equipment will entail the lapse of time
from the moment of decision to the moment of operation. This
period will have substantial costs. Thus, there would probably be a
step in the long run cost curve to F while the fixed costs expenditure on the new plant was made. Once the plant was in operation
there would be a dramatic reduction in variable costs. The relevant
cost curve is then TC_2. Secondly, in fact the enterprise will probably be unable to calculate exactly the value of the point B. It may
be preferable to make the investment too early rather than too
late, because if it is made too late some potential demand may be
turned away to rivals and be lost for a long period. For this reason
too the cost curve may be stepped as the firm moves to different
technologies.

From the long-run cost curve $ABCD$ may be derived the average
long-run cost curve of the firm in which technologies are not fixed.
The slope of the cord at G is the same as the slope of the cord at
H and hence the average costs are the same at outputs 6.5 and 11.
In between these outputs, however, the slope of the cord and
hence the average cost is much greater and hence the long-run
average cost curve has a hump over this range. This is shown in

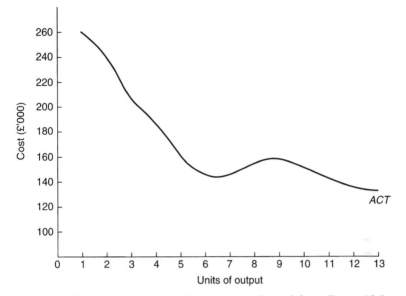

Figure 10.6 Long-run average total cost curves derived from Figure 10.5

Figure 10.6 – ignoring the effect of expenditure shown by *EFB* – for the portion of the total cost curve *ABC*.

There are, in practice, a very large number of indivisible items which cause discontinuous changes in fixed costs which, with non-linear variable total costs, can cause humps in the long-run average cost curve of the firm. At the very lowest level, a spade, for example, is not a useful implement at all if it is below a certain size, nor is a concrete-mixer or a crane. On the management side, the degree of specialised knowledge is such that it is barely possible to have a half-time project planner (unless his services are brought in as a consultant, in which case his cost will be high). Similarly, a research department has to be of a certain minimum size before it is a viable unit. Because of the 'lumpiness' of some factors of production, there will be certain stretches of the long-term cost curve which have some of the characteristics of the short-term cost curve. On a log scale of output it might conceptually look something like the diagram in Figure 10.7.

Clearly, as the output of the firm increases, the disrupting element would have to become larger and larger before it significantly affected the shape of the average cost curve, and on this diagrammatic presentation only a few of the possible changes have been

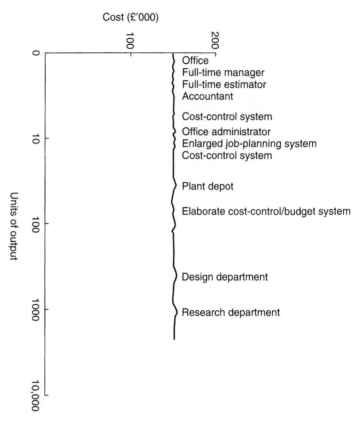

Figure 10.7   Possible humps in long-run average cost curves

shown. It was, however, deliberate to bunch some of these elements around 10 output units, because it is often considered by persons in the industry that there is a major point of crisis in a firm's growth, when it is too big for the directors of the firm to control each project without some further expertise, and too small to afford this expertise. This area is shown here around 10 output units. The real value of the 10 output units will differ according to the type of market the firm is operating in as well as to the calibre of the existing management.

Apart from indivisibility, about which there would be little argument, there is considerable disagreement as to the likely shape of the long-run cost curve. On the one hand, there are economies of scale *per se* which would tend to a falling cost curve, and on the other hand, there is supposedly one final indivisibility which is entrepreneurial ability (in which one might include not only the decision-making process but also the organisational framework) and also a possibility of rising factor costs.

Hague (1969) observes that such statistical evidence as there is suggests that in many firms the long-run cost curve does not turn up to the right but is more L-shaped, falling fast at first as fixed costs are spread over more and more output and then falling slowly or remaining more or less constant. He notes that statistical research has concentrated on rather large firms and says that it could be that in industries where small firms predominate, the long-run average cost curve slopes sharply up to the right. He goes on to say (pp. 117–19):

One reason why the long-run average cost curve slopes downwards is likely to be that there is technical progress. The firm can produce more cheaply as time goes on because its production methods become more efficient. This means that, strictly speaking, the curve that statisticians identify from looking at data over a period of time may not be a single curve but a composite one. For example, suppose that the firm in Figure 10.8* has experienced two big reductions in production costs during the time in question. It began producing in its present plant with the average cost curve AC_1; later costs fell so that the average cost curve became AC_2; a further fall in costs has meant that it now produces on the curve AC_3.

A statistical study might also show that in the initial situation the firm was producing the output OM with unit costs OC. Later, when the relevant cost curve was AC_2 the output of OM_1 might have been produced at a unit cost of OC_1. Recently the firm may have been observed to produce the output OM_2 at a unit cost of OC_2. It would be possible in a correct sense to join points A, B and C by a long-run cost curve. However, this would be a historical rather than an analytical cost curve. It would not, as the economist's cost curve usually does, show how cost and output were related in a given plant in a given state of technology. Before one draws conclusions about any individual cost curve identified by statisticians it is important to be quite clear whether it refers to one state of technology or to a number of successive states.

---

* The diagram number has been changed to correspond with those in this book.

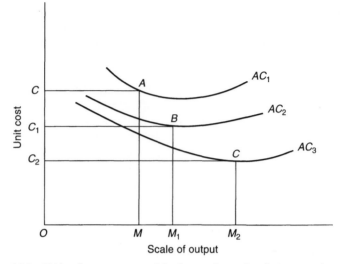

Figure 10.8   Shift of cost curves with change in technology over time

It is quite possible that it is the case in the construction industry that the large companies have been moving on a decreasing cost curve over time and through different technologies: for example, improvements in management techniques and the use of IT.

The advantages of being large are many, even once all the economies of using large units of specialised plant and knowledge have been obtained. First, a large company has advantages on the financial side. Its reserves for risk do not have to increase in proportion to its turnover as, with an increasing number of projects, the probability that they will all show losses together decreases. Another aspect of the decreasing risk with an increase in size is that the large firm can afford to take larger risks with a possible reward of higher profits on average. Moreover, if a company wishes to obtain finance, its size gives it direct access to the capital market and its very size gives confidence to investors.

One factor which is important in contracting is that the large company is sufficiently big to be able to obtain all the technical and managerial economies of scale in a number of markets, for example, housing and roads, at the same time, thus allowing it to gain the advantages of specialisation and of the risk-spreading of diversification.

On the employment side too, there are great advantages in being able to offer good career prospects for management, and in the purchase of materials, size may give the large firm bargaining advantages, not only in price, but also in delivery dates and service.

Much of the traditional economic analysis assumes that the long-run average cost curve of the firm will not go on decreasing or even stay at a constant level, but eventually must turn up. One reason why the firm may face a rising long-run average cost curve is that some inputs cannot be increased except at a higher price. This raises all sorts of questions about the level of competition which are discussed in Chapter 13. However, it is clear that if the firm has a large share of the market, the prices of its inputs could rise quite steeply. Sand and gravel, for example, may have to come from considerably further afield or from pits more expensive to operate if the demands are large. On the other hand it may be that the ability to purchase supplies of materials in bulk might offset any increase due to other factors.

A firm operating in a limited geographical area may well find that substantially to increase its turnover means extending its catchment area, and hence its costs of transport and supervision as travelling time increases. Similarly, the geographical spread of demand for projects may not coincide with the availability of manpower so that the price of labour may rise with expansion in turnover, either because of increased transport costs or because of the bargaining power of the operatives.

Another reason put forward for eventually increasing long-run average costs is the indivisible nature of 'entrepreneurial ability' – that the decision-making process gets clogged. Williamson (1967) has applied organisation theory to this problem. He says:

> For any given span of control (together with a specification of the state of technology, internal experience, etc.) an irreducible minimum degree of control loss results from the simple serial reproduction distortion that occurs in communicating across successive hierarchical levels. If in addition goals differ between hierarchical levels, the loss of control can be extensive.

There may be many firms which are on a long-run increasing cost curve. Only detailed analysis in the field (in which they would probably be unwilling to cooperate) would show whether the reason for their being at their existing size is that they are in a situation of rising costs. In the world as a whole, writing in the year 1999, it seems firms are still trying to grow, often by linking up with other firms.

Problem of growth over time

Some reference has already been made to the problem of assessing costs over time when technology has changed (see quotation from Hague (1969) p. 119). In addition, in a consideration of short-run costs it was found that there were costs of changing from one level of output and, in the instance quoted, particularly from one level of employment to another and that these might not be the same when output was rising as when output was falling; that is, the cost curves worked in one direction but were not reversible.

The problem of time and of rate of growth over time affects a great many other relationships between output and cost. First, there are advantages, not only in being large as already discussed, but also in *getting* large. It is generally easier to introduce technological change if, at least for the pilot project using the new process, skills can be brought in from outside without detriment to the employment of those already in the organisation. For a factory process it is easier to open a factory using new methods and a reduced *proportion* of manpower if the growth of the firm is such that the total men employed need not be reduced, thus leading to poor morale. Similarly, a growing company – even more obviously than a large company – will be able to attract high-calibre staff at a relatively favourable price.

There are corresponding difficulties and costs for a company which is shrinking in size: the UK redundancy legislation mentioned earlier is just one of the many costs. However, the adjustments which must take place in the organisation of the firm as it grows and the problems of adjustment mean that there is probably a maximum rate over time at which growth can take place or, to put it in theoretical terms, there is a maximum rate at which the factors fixed in the short term can be made variable. In theory, it might be possible to change from a firm of £100,000 turnover to £100 million turnover in the space of time in which factors fixed in the short run could be varied. In practice, it seems necessary to go relatively slowly through the in-between stages (partly because of the problem of finding work) because of the other factors not thought to be fixed, which are discovered to be fixed in practice. One reason, for example, why a firm may find it difficult to grow at a fast rate is that working capital will not be available, i.e. it is in danger of overtrading. If it is able to obtain capital, it may have to pay a higher rate of interest. This is not a cost in the usual sense if the

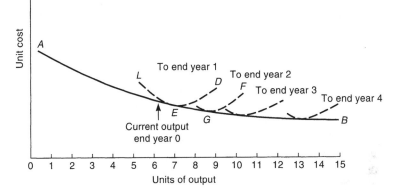

Figure 10.9 Average cost curves for each of a series of years, assuming two-year planning periods

willingness of banks and others to finance the firm is determined more by the *rate* at which they are willing to increase lending than by the absolute level of lending.

It is suggested therefore that there will be a cost curve, which is valid for a planning period, which rises as output is increased beyond a certain level because of the difficulty of absorption of the additional resources needed, particularly in terms of human relationships and appropriate organisational adjustments, and also possibly in the difficulty of obtaining the resources of working capital. If the latter is obtained by a bank loan, for example, its provision will be dependent on satisfactory progress at recent rates of growth, thus precluding a very rapid increase in this rate. Figure 10.9 shows such a series of curves, on the assumption that the long-run cost curve *AB* is declining slightly over a long range of output and that management, labour, plant and equipment, etc., can be obtained in as large quantities as required. Each cost curve is considered to be relevant to a year, but the planning period is assumed to be longer – say two years. Thus at the end of year 1 plans can be made of increasing output in two years' time to 7 units, i.e. to *E*. However, if output is pushed beyond *E* in this period, costs will rise substantially along *ED*. Assuming that output will be satisfactorily extended to *E* at the end of year 1, at the end of year 0 plans can be made for year 2 for which the effective cost curve (by the end of year 1) will be *EGF*. However, if plans are not made two years in advance, the cost curve for year 2 would probably turn up before point *G*. The increase in output possible in each succeeding period will not always be constant because, as indicated by Figure 10.7 for example, the problems will be different at each stage.

These period cost curves can be regarded as special cases of short-run cost curves over time. Although in the exposition above it has been assumed that the firm is aiming at growth, in fact it is equally true that if it falls behind its planned expansion it is also likely to experience an increase in costs, as shown by the U shape of the curve, for example *LED*.

It is difficult to substantiate this theoretical concept from recent data on firms. Studies of the large firms in UK industry (Hillebrandt and Cannon, 1990; Hillebrandt *et al.*, 1995) of the boom period of the 1980s and the 'fall' of the early 1990s find the industry almost entirely attributing the fall in profits to demand factors, although they should have anticipated the fall in demand, and therefore management is to some extent at fault.

It is quite frequent in discussions with senior management in the industry for their plans to be expressed in terms of a desirable rate of growth in output. As has been mentioned earlier, it is not easy in the construction industry, by reason of the uncertainties of the tendering situation, to regulate work load. Sometimes a firm wins more tenders than it anticipates. Once a firm has acquired the work and recruited the necessary staff, it is loath to dismiss it again and revert to the planned growth rate. This staff tends to become the new norm, and planned growth is superimposed on the new higher norm. If this is correct, it would account for the belief among observers of the industry that many of the failures are basically due to lack of working capital.

Supply curve of the firm

The output which the firm is prepared to supply at any price is shown by the marginal cost curve so long as this lies above the average cost curve. This must be so, for if the firm is not covering the additional cost of producing another unit of output, then it would be better not to produce it. When the marginal cost curve lies between the average total cost curve and the average variable cost curve, the firm will still find it worthwhile to produce near the output shown by the marginal curve. This is because in this area – small for most contracting firms because of the low level of fixed costs – it is covering the additional costs of the project even though it is not making its full contribution to head office overheads. If the marginal cost curve is below the average variable cost curve, the firm should not produce at all. In all this analysis the relevant marginal and average cost curves are those related to whatever time period is under consideration. Thus, in the short run the supply curve of the firm will slope upwards to the right fairly steeply. In the long run, the supply curve may be relatively constant for quite large ranges of output, although, assuming unchanged technologies, it would probably eventually turn up.

11
Market Supply Curves

While it is a truism that the supply curve of the market is made up of the total of what the firms in that market are able to supply at various prices, the variability of firms complicates the summation of individual supply curves. Moreover, other factors, which could be ignored for the firm having a relatively small share of the market, may dominate the market supply curve, thus making the total market supply curve different from the sum of the individual firms' supply curves. These problems are discussed below and the conclusion for the supply curves of the firm and of the market are summarised in Table 11.1.

Efficiency of firms

If firms were all of equal efficiency, there were no changes in price of factors of production with changes in output, and firms were all aiming to maximise profits, the supply curve of the market would be the same shape as the supply curve of the individual firm (although with different values of output on the horizontal axis). The position is illustrated in Figure 11.1 where *AB* is both the supply curve of the individual firm, using the firm scale of output, and of the market, using the market scale of output and assuming that there are 100 firms operating in this market.

If, however, the firms are different in efficiency owing to differences in entrepreneurs (the other assumptions remaining the same), the most inefficient producer will have the highest average cost curve – and it will be just covered by price, otherwise he will leave the industry – and the other producers will have lower average cost curves which will be more than covered by price so that they

Table 11.1 Possible variants in the supply curves

	Short run for firm	Long run for firm	Long run for firm and market
INDIVIDUAL FIRMS			
1. Average costs:			
(a) increasing	P	P	P
(b) constant	–	P	P
(c) decreasing	–	P	P
2. Importance of profit motive:			
(a) firms maximise profits	P	P	P
(b) firms do not maximise profits	P	P	P
3. Time-span:			
(a) short run – some factors fixed	P	–	–
(b) long run – no factors fixed	–	P	P
(c) unlimited amount of growth possible	–	–	–
MARKET			
4. Degree of homogeneity:			
(a) all firms equal efficiency	P	P	P
(b) firms not equal efficiency	P	P	P
5. Extent of shortage of factors of production:			
(a) some factors in short supply	P	P	–
(b) no factors in short supply	–	P	P
6. Prices of factors of production:			
(a) increasing with increase in demand	P	P	P
(b) constant with increase in demand	P	P	P
(c) decreasing with increase in demand	–	P	P
7. Number of firms:			
(a) fixed	P	P	–
(b) not fixed, i.e. free entry and exit	–	P	P

P = possible situation.

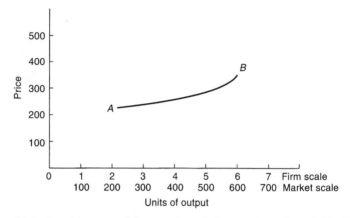

Figure 11.1 Supply curve of firm and market on assumption of identical cost curves of firms

make higher profits as a reward for their superior entrepreneurial ability. These profits may or may not be higher than normal profit, which is just sufficient to keep them in the industry. Whether or not the market cost curve with variably efficient producers will be higher or lower than that with homogeneous producers cannot be known, since the efficiency of the homogeneous producers relative to the heterogeneous ones is not known.

In fact, it is clear that in the construction industry all entrepreneurs are not equally efficient and therefore that some producers have lower costs than others and may earn profits rather higher than those which would be just enough to keep them in the industry.

The general principle of the summation of the cost curves of the individual firms to obtain the supply curve of the market may be applied for the short run or for the long run.

Importance of profit objective

It was suggested under the consideration of the objectives of the firm that firms might not all maximise profits or wish to do so. In this case, the effective supply curves of the firms which have to be summed are not the theoretical cost curves showing how much they can economically produce at a given price, but the effective cost curves showing how much they are actually *prepared* to produce at a given price. For example, it may be that family firms wish to limit the size of the firm to that which can be managed by the family, or that after a certain amount of profit the entrepreneur values leisure more than profits. In both these cases their supply curve will be different from their marginal cost curve, with a cut-off point more like that in manufacturing industry due to fixed plant. Thus, the total effective supply at any price will be lower if some firms are deliberately restricting output than if all were prepared to supply according to the objective of maximum profits.

Time span

The usual conceptions of short run and long run have been dealt with in discussion of cost curves of the firm. It was, however, suggested that in addition there were period cost curves; that is, that there was a maximum rate at which a firm could expand. This concept of the period cost curve of the firm is a further complication in the adding-up problem, because it means that the traditional long-run cost curve of the firm is summable for all firms only over a period of time much longer than normally considered.

Shortage and price of factors of production

It is quite possible that the market will be unable to obtain certain factors of production in the quantities required: for example, there simply may not be enough of a certain type of engineer to manage a large increase in the volume of a type of specialised project; there may not be enough carpenters in a certain area to enable the traditional housebuilding programme to be, say, doubled; there may not be enough copper pipe to allow a substantial increase in the rate of output of domestic central heating systems. Some of these may create a definite limit for a period of time to the output within the market, but it will probably at the same time create a change in the cost curve of the market, because the increased demand for a particular commodity will tend to raise its price to a new equilibrium balancing its demand and supply.

Other shortages of factors of production may be able to be overcome in physical terms, but again only by an increase in the cost curve of the market. It may be possible to attract specialist engineers from abroad; to pay overtime to carpenters or provide transport and additional earnings to bring them from other areas; to import copper pipe from abroad. Some of these methods of overcoming the shortage will provide equally efficient resources at a higher price, others may provide less efficient resources at the same price, and others again less efficient resources at a higher price. In each case the effect is to raise the cost curve of the market.

In the long run, the industries (or other institutions) producing the factors of production will react to the shift in the demand curve for their product and will be able to expand output. It will depend on the cost curves for the factors of production whether the long-run price will be higher than, the same as or lower than at lower outputs.

There will be no increase in the cost of factors of production (or a shortage of supply) if an individual firm alone expands its output, so long as its market share is imperceptible. However, if there are a few firms in a market, any one firm may be affected by a shortage of factors of production. In this connection it should, however, be pointed out that any market for the products of the construction industry may be served by factor markets which cover a number of different markets, e.g. steel, so that, except in specific cases, of which some examples have been suggested above, the individual firm and the individual market may have a relatively small share of the total demand of factors of production.

Entry to and exit from the market

One of the conditions of equilibrium in the industry is that there is no incentive for firms to enter or leave the industry.

A firm will in the long run leave the industry if it is not making normal profits. It needs to cover the rate of return the firm could attract on its capital elsewhere, plus remuneration for its directors equivalent to that which they could earn elsewhere, adjusted for the entrepreneurs who are content to accept a lower rate return on themselves and their capital because they value other advantages more highly. These entrepreneurs may or may not be maximising profits.

Firms will wish to enter the market if they think they can make at least normal profits in the market, either because they think they are more efficient than the least efficient producer, or because the ruling price is higher than the average cost of the least efficient producer – owing, for example, to a shift in the demand curve. The latter situation may obtain, even if the long-run cost curve of the existing firms is constant or decreasing, if the rate at which the existing firms can grow is lower than the required increase in demand in the time period necessary for new firms to enter the market.

In the long run, free entry to and exit from the market will tend to lead to a situation where all firms in the market are producing at minimum average cost so that the long-run market supply curve may be horizontal, or rising slightly if new firms are less efficient.

Discussion and assessment of factors affecting the market supply curve

It will be seen from Table 11.1 that in each period considered the possible combinations of each of the variants is very great. Even if some of the variants are assumed to be constant, for example the average costs, it still leaves a number so large that it is neither feasible nor helpful to construct a model for each situation. Some general observations may, however, be made.

The only situation in which factors not directly considered in the cost curve of the individual firm are likely to lead to a decrease in the long-run supply curve of the market is that in which factors of production have a decreasing cost curve, that is 6(c) in Table 11.1. Although this may be the case for one or two inputs of the construction market, it is unlikely to be true over the whole range of inputs. Therefore, having regard to the fact that at any given level of knowledge the long-run average cost curve of the individual firm is likely to be decreasing at best almost insignificantly, it is virtually ruled out that the long-run supply curve of the market at any given time is other than rising, although over a long period of time with different technologies it could slope downwards (See Chapter 10).

The short-run supply curve of the market will also be rising and will, as in the case of the individual firms, be considerably steeper than the long-run supply curve.

12
Equilibrium in Various Market Situations

It was shown at the beginning of Chapter 10 how the costs of the individual large contract, obtained at a single point in time but with work spread over a long period, are relevant to the usual cost curves of economic analysis which represent the answer to the question: If the output of the firm were higher or lower than a given level, what would be the effect on costs? The remainder of the chapter was devoted to a detailed consideration of this question.

By an analysis almost identical to that for costs, the revenue obtained from individual projects may be transformed into revenue over time, and then consideration be given to the likely difference in revenue at a point in time if the output had been larger or smaller than a given level. This is the demand curve of the firm. It is dependent, on the one hand, on the demand curve of the industry or market under consideration and, on the other hand, on the position of the individual firm in the market in question. The first was discussed in the demand section of this book, in Chapters 4–8. The second requires a preliminary general investigation of the various theoretical types of market situation, because the demand curve of the individual firm in some conditions, notably oligopoly and discriminating monopoly, cannot be understood without reference to the analysis of both demand and supply. This is the subject of the present chapter. In Chapter 13 consideration is given to the place of the construction firm in various markets on the theoretical range from perfect competition to monopoly.

Before embarking on the analysis of cost and revenue in various market situations, it is advisable first to pause and consider the relationship of average revenue, which is the demand curve facing the firm or the industry, to total revenue and marginal revenue, both of which concepts are necessary for the analysis following.

Average, total and marginal revenue

The demand curve is the average revenue curve of the firm or the industry as the case may be. From it the total revenue and thence the marginal revenue curve may be derived. Alternatively, just as in the case of costs, the analysis may commence with the total revenue of the firm and the average revenue be derived from it by the slope of the cord and the marginal revenue by the slope of the tangent. The pairs of diagrams in Figure 12.1A–E show the relationship between these in varying demand circumstances.

In Figure 12.1A, elasticity of demand is less than 1, which means that following the analysis in Chapter 4, the total amount spent on the commodity decreases and, as elasticity is constant, the total revenue curve is a straight line. Clearly, the marginal revenue must be constant and negative because the tangent to the falling revenue curve is constant and of negative slope. The market demand for the output of construction is usually inelastic.

Figure 12.1B shows elasticity of demand equal to unity. This means that the total amount spent on the commodity is constant and the average revenue curve is a rectangular hyperbola. Marginal revenue is nil and constant.

Figure 12.1C shows elasticity of demand greater than 1 and constant. The total revenue curve is a rising straight line, the average revenue is a rather flat curve and marginal revenue is constant and positive.

Figure 12.1D shows the situation where the elasticity of demand is infinite. This is of interest because it represents the position of the demand curve of the firm under perfect competition. Average revenue equals marginal revenue and the total curve is a straight line with a slope of 1.

Lastly, in Figure 12.1E the elasticity of demand is first greater than 1, then less than 1. The demand curve, i.e. the average revenue curve, is shown in this diagram as a straight line falling to the right, and the marginal revenue curve similarly decreases to the right. From this diagram it can be seen that the average revenue curve is the slope of the cord of the total curve and the marginal revenue curve the slope of the tangent. When marginal revenue is nil, the total is at its highest point.

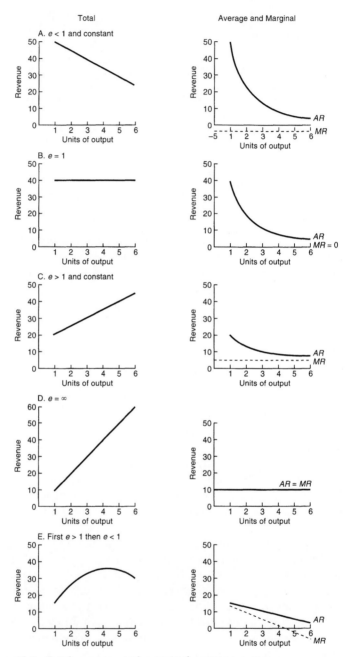

Figure 12.1   Total, average and marginal revenue curves

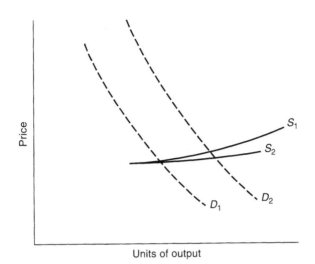

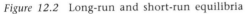

Figure 12.2   Long-run and short-run equilibria

Market equilibrium

Equilibrium exists in the market when:

(*a*) the price is such that demand and supply are equal at that price;

(*b*) there is no incentive for firms to move out of or into the market;

(*c*) there is no incentive for producers to change their output, method of production or price.

In all markets the price at which goods will be sold is that which balances the amount on offer with that demanded. It was found in Chapters 4–7 that the demand curve for construction will have varying elasticities depending, *inter alia*, on the type of construction under consideration, and in Chapter 11 the short-run supply curve was found to slope upwards to the right and the long-run supply curve to be gently upward-sloping or even horizontal. Over a long term in which technical knowledge is not fixed, there may well be a long-run cost curve for the industry which decreases over time. Thus, while in the short run a shift in the demand curve for construction may well cause a rise in price, in the long run the industry may be able to meet demand with little or no increase in costs. This is illustrated in Figure 12.2, in which D_1 is the demand curve in period 1 and D_2 is the demand curve for period 2 shifted upwards, as a result, for instance, of a rise in real incomes. S_1 is the short-run supply curve which shows that in the short-term the new demand will be met only at a higher price. S_2 is the long-run supply curve more or less constant over a large range of output, indicating that eventually the industry will be able to adjust to the changed demand situation.

In practice, an equilibrium position is never (or very rarely) reached because circumstances are altering all the time. In particular, the demand curve will shift as tastes, incomes, expectations and so on all change marginally over time. The market will, however, always be attempting to move towards an equilibrium position. To understand the changes in the market supply position, it is necessary to investigate the various possible relationships of the firms in the market to each other and to the total market situation.

Economic theory distinguishes a number of market types with perfect competition at one extreme and monopoly at the other extreme. In between is an infinite range of imperfect competition. On the buying side too there is a range of possibilities from one buyer – monopsony – to an infinite number of buyers.

Perfect competition

The main characteristics of a perfectly competitive market are:
1 Homogeneous product; that is, each producer must be selling a product which is indistinguishable from those of other producers.
2 Large number of firms so that each firm produces an insignificant proportion of total output.
3 Free entry to the market.
4 Perfect knowledge of what everyone else is selling and at what price.

Similar criteria exist on the purchaser's side – in particular that there are so many buyers that each one purchases an insignificant proportion of total output.

The existence of all these conditions together leads to a condition of perfect competition – sometimes approached, for example, in a grain market, but never actually in practice achieved. Sometimes a rather more realistic model of pure competition is used embodying criteria 1–3 but dropping that of perfect knowledge.

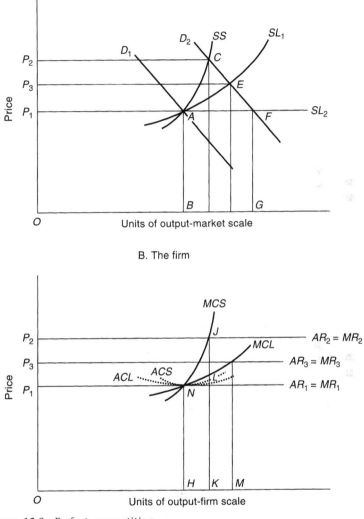

Figure 12.3 Perfect competition

In perfect competition in equilibrium all firms will have the same costs. This does not mean that all firms are identical, but simply that if one firm has, say, a more efficient manager than others, then the price which he will be able to charge for his services will be such that he takes all the benefit he gives the firm with his superior efficiency. If he were not paid this amount, under competitive conditions he could move to another firm where he would be.

In Figure 12.3A, SL_1 and SS respectively are the long-term and short-term supply curves of the market assuming that no new firms enter the market. They have the same shape as the relevant part of the marginal cost curves of the firm in Figure 12.3B. In Figure 12.3A D_1 is the demand curve. It cuts the supply curves at A so that output is OB and price OP_1.

Assume then that there is a shift in the demand curve to D_2 in Figure 12.3A. The short-run equilibrium will be at C with price P_2. In the long term, if there were no possibility of other firms entering the industry, equilibrium would be at E with price P_3.

The position of the firm is shown in Figure 12.3B. ACS and ACL are the short and long-run average cost curves and MCS and MCL the marginal cost curves of the firm. AR_1 is the average revenue curve at the market equilibrium price P_1. This is the price which the individual firm obtains no matter how much it produces. The output of this one firm is, according to the definition of perfect competition, such a small fraction of the market's output that even a large change in the output of this firm affects the total output of the market to only an infinitesimal degree. If the firm sells at a price below this ruling price, since there is perfect knowledge, it would be able to sell all its output but its average costs would be greater than average revenue and it would therefore go out of business. If it quoted a price above the ruling price level, it would sell nothing because, again assuming perfect knowledge, every buyer would go elsewhere. As the average revenue is constant, it is equal to marginal revenue so that AR_1 is both the average and the marginal revenue curve. The optimum position for any firm is always that at which its marginal revenue equals its marginal cost. This must be so because, if the revenue from an additional unit of output were greater than the cost of producing that additional output, then it would pay to go on expanding output until there was no advantage in further expansion, i.e. so that marginal revenue was equal to the marginal cost of its production. Similarly, if the revenue from an additional unit of output is less than the marginal cost of its production, then the firm would be better off contracting production until it is no longer making a loss on the last unit.

Thus, at price P_1 the optimum position for the firm is at the output at which the marginal revenue curve MR_1 cuts the marginal costs curves MCS and MCL, i.e. at N where output will be OH. The average cost and average revenue at this point are also equal, so that the firm will earn only normal profits. When the market demand curve shifts, the new short-run price level ruling in the market becomes P_2. The new average and marginal revenue curves of the firm become AR_2 and MR_2. The optimum output becomes OK, i.e. that output at which the marginal revenue curve MR_2 cuts the marginal cost curve MCS at J. At this output the average revenue or price is KJ and the average cost KL, so that the firm is making a supernormal profit of JL on each unit of output. In the longer term, but with the number of firms still fixed, the price becomes P_3 and output rises to OM. The firm still has a supernormal profit.

If new firms can enter the industry, they will do so because they will be attracted by the supernormal profits. If they can obtain factors of production at the same prices, they will enter until they shift the long-run supply curve to SL_2 in Figure 12.3A. Equilibrium will then be at F with price P_1 as before and output OG. The output of the firm will revert to its original level at output OH in Figure 12.3B with no supernormal profit. The increase in the output of the market as a whole would then be made up entirely by the output of the new firms entering the market.

Monopoly

At the other extreme of market conditions is monopoly, i.e. where there is one producer and seller. In these circumstances, the average revenue curve of the monopolist is the same as the demand curve of the industry, so that the average price received per unit of output falls as more output is put on the market. When average revenue is falling, marginal revenue is below average revenue and falling faster. Thus, in Figure 12.4, if AR is the market demand curve and the average revenue curve of the monopolist, then MR is the marginal revenue curve. AC is the long-run cost curve which might approximate to the average cost curve of the industry under competitive conditions. On the one hand, it might be lower because the monopolist might be able to obtain economies of scale by unifying production and, on the other hand, it might be higher as the

140   *The Supply of Construction*

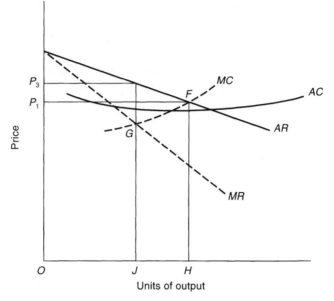

Figure 12.4 Monopoly: the market and the firm

monopolist might require a more bureaucratic organisation and, for example, indulge in a large industry advertising campaign. The marginal cost curve is *MC*. The price under competitive conditions would be determined by where the demand curve cuts the supply curve, i.e. the marginal cost curve, at *F* with price P_1 and output *OH*. In the case of the monopoly firm, however, the position of maximum profit would be obtained where the marginal cost curve cuts the marginal revenue curve, i.e. at *G*. Price would be P_3 for output *OJ*, i.e. a higher price and a lower output than that under conditions of competition.

It is possible for a situation of monopoly to be one of equilibrium according to the three conditions listed above. The price is one which equates demand and supply – although the supply is entirely determined by the one producer and there may well be no incentive for the producer to change his output, method of production or price. The real question arises on whether there is incentive for firms to move out of or into the market. Situations do arise in which the monopoly is based on technical economies of scale. If the cheapest way of producing the product is on a scale which is as great as the whole industry, there would be no incentive for any firm to enter the industry for it could not compete at an output lower than that of the whole industry. (Indeed, it would be theoretically feasible that the lowest cost of the optimum scale of production was achieved at outputs greater than the industry output. In this case the cost curve could be falling, and if it were steeper than the demand curve there would be no equilibrium position). In other cases, however, the monopolist is being constantly threatened by new entrants and he may have to limit the amount of monopoly profit he makes in order to discourage new entrants. In practice, he may find it desirable to limit profits and behave in a way similar to the competitive situation, so that he does not break any anti-monopoly law or lead to a situation in which legislation would be considered desirable.

One peculiar form of monopolistic situation must be mentioned, namely that of discriminating monopoly. This arises when there is no possibility of reselling the product from one buyer to another. This enables the monopolist to split up his markets and sell at a higher price in markets where demand is relatively inelastic and at a lower price in markets where demand is elastic, thus maximising profits in each market. The sum of the profits will be greater than the profit obtained by treating the market as one. The classic example of this practice is that of the doctors who charge higher fees to rich patients than to poor ones. A limited form of discriminating monopoly may also exist when the product of the market may be resold but where the resale value is for some reason lower than the value at the point of first sale. The latter case has some relevance to construction, as will be seen later.

Monopsony

Monopsony is the situation in which there is a single buyer. Because construction products are custom-made and no one else is likely to want that same product, nearly every buyer of the construction industry's output is a monopsonist. However, if the construction industry's output is regarded, not as products, but as the service to create buildings and works, there is not one buyer but many. The significance of monopsony in construction needs to be considered in relation to the specific market situation.

Other forms of imperfect competition

In fact most industries and markets lie somewhere in between perfect competition and monopoly. It will be convenient to consider these in two main sectors of the number of firms and the extent to which products are differentiated.

Oligopoly

A market in which there are few producers and in which, therefore, the actions of the individual firm do have an effect on the overall market is known as a situation of oligopoly, i.e. a few producers, just as monopoly denotes one producer. Each of the firms in a position of oligopoly will have a share of the total market. Assuming the market average cost curve is the same for oligopoly as for monopoly, the optimum price, that is, that which maximises profits, at which they would sell their products would be the same as that at which the monopolist would sell, as this is the price which maximises profit in the market as a whole and would therefore maximise the size of the share of each oligopolist. However, if one or more of the oligopolists decided to cut price to obtain a larger share of the market, because there were so few producers, his action would be obvious and would be seen as a threat to the market share of the others. They would therefore tend to cut their prices by the same amount or more. This process could be continued until the oligopolists had so reduced their prices that they were earning only normal profit, and in this case their output would be the same as that under perfect competition. The actual price level could be anywhere in between these two extremes.

It is interesting that, in a position of oligopoly, there might well be no reaction from competitors to a rise in price. They would not see this as a threat to their positions. Thus the demand curve facing the oligopolist may well be kinked, as shown in Figure 12.5. Assume the ruling price is that at point *B* on the demand curve *AD*. Then, above that price, if the oligopolist increases price this will not induce his competitors to act similarly and he moves along the curve *AB* – which is essentially his share of the market demand curve. If, however, he reduces price, the other producers will also reduce price so that his output will not increase as much as if he could do this without retaliation and move downwards along his share of the market demand curve from *B* to *D*. In fact, he will find himself on the line *BC*. His average revenue curve will therefore be

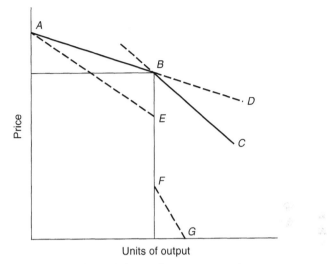

Figure 12.5 Oligopoly: kinked demand curve

ABC. The corresponding marginal revenue curve is the broken line *AEFG*. If his marginal cost curve passes between *E* and *F*, then the entrepreneur will not move from his initial position at *B*. Not until his marginal costs are as low as or lower than *F* would it pay him to reduce price. For this reason, in a situation of oligopoly, there is a strong incentive for all the firms to form a tacit or overt agreement not to enter into a price war. Competition may tend to take other forms, such as that of product differentiation.

Product differentiation

In a market in which there are many firms, each producing a product slightly different from that of its competitors, each firm will be faced by an average revenue curve showing the demand for its own products which will be nearly horizontal, that is, near to perfectly competitive conditions, if its own products have very close substitutes in the market as a whole. It will slope the more, the greater the degree to which the firm is able to convince buyers that its products have special features. These features in construction could include reputation for quality workmanship or special skills, accessibility of senior manager to the client, suitable location of national or regional offices and, for work abroad, appropriate local partners and language abilities of staff. Other, more costly, strategies include taking equity in the project, arranging a financial package, offering a warranty for the building upon completion or securing a key tenant. This type of market is known as monopolistic competition.

Product differentiation, may also occur in markets where there are few producers, a situation of oligopoly. In this case the oligopolist will be able to make price changes without so much likelihood of retaliation by his competitors. Price competition, at least, is likely to be less severe than in oligopoly with homogeneous products.

Measures of the extent of monopoly power

It will be noted from the above discussion of various market situations that the more the individual producer is able, through product differentiation or simply by being one of a few in the market, to keep to himself a piece of that market, the greater the slope of his average revenue curve (assuming that the market demand curves have equal elasticity). Thus, a measure of the degree of monopoly power is the difference between the price which the firm charges and its marginal cost. In perfect competition, marginal cost is equal to price. In monopoly it is considerably less than price. This measure has disadvantages, however, since it is dependent not only on the market form, but also on the slope of the industry demand curve. Another measure of the degree of monopoly is the extent of supernormal profits earned by firms in the market. Clearly, both measures have value, but the overall situation must be examined before a judgement is made.

Contestable markets

Freedom of entry has already been mentioned as a condition of perfect competition. It is also the principal defining characteristic of another type of market situation, which can apply to any of the types in the range from perfect competition to monopoly, namely a contestable market. Willig (1987) describes contestable markets as 'those in which competitive pressures from potential entrants exercise strong constraints on the behaviour of incumbent suppliers. For a market to be contestable, there must be no significant entry barriers.' He goes on to say that this applies to a monopoly situation or one with a large number of competing firms, because it is dependent on potential competition from possible entrants, rather than merely on competition among active existing suppliers.

Thus, the idea of contestable markets takes one of the main conditions of perfect competition and elevates it in importance to apply to a number of different market situations in a way that is in keeping with the practical working of some markets, in contrast to the idea of perfect competition which, although useful to define a theoretical limit, is never fully realised in the real world.

The concept of the contestable market was developed by Baumol *et al.* (1982) in a book which was to colour the thinking on market structure of economists and policy makers alike. In particular, it has important lessons for the regulation of monopolies and clarifies the circumstances in which monopoly is benign.

Objectives other than profit maximisation

The effect of assumptions other than that the entrepreneur wishes to maximise profits must be considered. First, the situation of perfect competition will be examined. In this situation each firm is producing at the lowest part of its average cost curve and making only normal profit. If the entrepreneur wished to limit or increase the output of the firm, he would move either back and upwards or forward and upwards on his average cost curve, which, with a constant average revenue curve, means that his profits would be reduced below the normal level. If, on the other hand, he continues to obtain normal profits, then the secondary goal of revenue maximisation implies the same output, because in perfect competition with homogeneous products, average revenue equals marginal revenue equals average cost equals marginal cost. Thus if he wishes to maximise profits he produces the output at which marginal revenue equals marginal cost, which is the same output as that for revenue maximisation with normal profit, the situation where average revenue equals average cost.

In conditions of monopoly the entrepreneur is able to restrict output so that he earns less than maximum profit but his profit is still greater than or equal to normal. If he wishes to earn normal profit and maximise revenue he will extend his output until his average costs (including normal profit) equal his average revenue. Thus, in Figure 12.4, if the entrepreneur did not wish to earn more than normal profit but wished to maximise revenue, he could move from the profit-maximisation output of *OJ* to the revenue-maximisation output of *OH*. This is also the output which would be produced in a perfectly competitive market.

13
Demand Curves Facing the Individual Firm

In discussing the demand curves facing the firm, it is important to realise the diversity of practice across the world. In a substantial number of countries, open or selective tendering is the procurement method for most, often virtually all, of the projects in the formal organised sector. In the description of how work is put to the industry in Chapter 8, about ten alternative procurement methods were described and there are, in fact, many variants of these. However, even in the UK, where some of them have been developed, they are still responsible for procurement in only a relatively small percentage of total construction projects, although, because the projects are generally large, they account for a high percentage of output. This is because many of these procurement methods are not suited either to small projects or to repair and maintenance which, in developed countries, account for a high proportion of the work. In the discussion which follows, attention is given to these unusual projects both because they pose interesting questions for the future shape of the construction industry and because their use is spreading and is likely to spread further.

Having defined in Chapter 12 various theoretical types of market situation, it must now be determined to which broad types construction markets belong. First, some of the determining factors are considered specifically in relation to construction.

Homogeneity of product

Bearing in mind that the product of the construction industry is a service, even within the individual construction market, that product is not completely homogeneous. Indeed, in all cases of price determination except open competitive tender, it is in the interests of the contractor to convince the client that the service he has to offer is in some way superior to the service offered by his competitors. He hopes that the client might choose him to be on the selective tender list or to negotiate a contract. Indeed, the contractor wishes to move as far as possible away from the standardised product in the perfectly competitive market to imperfect competition with its product differentiation.

Under the traditional system, attempts at product differentiation normally all cease when the project gets to the tendering stage because, at this stage, it is assumed that the products are homogeneous and that competition takes place on price alone. Since it is assumed that at the tendering stage the products are all the same, the market on this criterion is near to perfect competition.

The situation is very different for the less traditional types of procurement. By their very nature, the product cannot be the same for all competitors. One of the objectives of design and build or BOOT, for example, is to see what different solutions the contractor or other team leader produces. None of the less traditional procurement methods has homogeneous products.

Number of firms

Within some of the construction markets there are so many firms that each produces an insignificant part of the total output. This situation is most likely when the market is large and when the size of the firm is small, for example, in general contracting on small to medium-size work without any complex technical problems in a metropolitan area. In such cases, there are sufficient firms to lead to effective competition at the tender stage. This means that at this stage the client has a large choice of contractors from which to select for tendering or negotiation.

In many countries, developed and developing, there are contractor registration schemes under which a contractor is registered as being capable of undertaking work up to a certain size and of a stated type. In the UK, there are some schemes by which contractors register with a client body on a purely voluntary basis, but if they are not registered, they are unlikely to be considered for work. The Design Build Foundation, for example, is setting up a registration scheme of participants in the process from which members will select contractors, consultants and specialists for their projects.

Also in the UK, a voluntary registration scheme for small builders is proposed, to overcome the poor quality service given by so called 'cowboy builders'. The objective of these schemes is to simplify the selection process and/or to increase the quality of construction, but it also reduces the number of firms in any particular market. Whether a registration scheme will reduce the number of firms in the market so severely as to limit competition, depends on the number of firms remaining in the market and other circumstances, such as location. There is no doubt that they tend to limit free competition, although they may have positive advantages which outweigh the disadvantage of reduced competition.

In markets which are small, either because they are limited geographically, such as a small market town in an otherwise rural area, or because they are limited by the degree of specialisation, for example, suspended-ceiling specialists, or both, there may well exist a situation in which there are only a handful of contractors in the market for any particular size and type of job. This is particularly likely to be so in new types of work, such as multi-storey flats or motorways in the immediate post-war period. Clients do not then have a sufficiently large choice of contractors to fulfil the requirements of an adequate number of firms for anything approaching perfect competition.

In a situation of open competition, the number of firms who tender is potentially the whole of the market for that type of contract, but many contractors have decided that they will not do work for which selection is simply by open competition; in effect, therefore, they have created another split in markets, by method of appointing the contractor. Within this market it is almost a tautology to say that the firms entering, constitute all the firms in this market at that time. The extent of competition, as far as the number of firms is concerned, is the same as the extent of competition in the market as a whole and, as has just been found, this ranges from effective competition, to something which may be considerably less.

In cases of selective tendering, however, the number of firms who participate in the process of price determination, and hence compete on price, is by the very nature of the process deliberately reduced, so that they are less than in that market as a whole. There are a number of codes of practice for selective tendering. If the recommendations of the Code of Procedure for Tendering by the National Joint Consultative Council (NJCC) (1994) are followed, the number will range from five to eight and will certainly be so few that the price submitted by any one will take into consideration the likely actions of his competitors, so that it becomes a situation of oligopoly. The ways in which he can, and to some extent does, take his competitors' position into account is dealt with in detail in Chapter 14 on tendering.

If the number of contractors within any one market falls below a certain level, it becomes more feasible to contemplate some sort of agreement to divide the market so that market shares are predetermined, or to fix price levels. In such a case, the group of contractors are virtually, by concerted action, bringing themselves nearer to the extreme case of monopoly. There have been situations in the construction industry where some form of agreement is alleged to have been operated, and there may still be some. Apart from the restrictions of law, however, the durability of such agreements is severely limited by the free entry to the industry and its markets, dealt with below.

Another way in which competition is limited at the tendering stage is by the taking of cover prices by a contractor not wishing to obtain the job, but unwilling to tell the client that he would rather not tender. In such a case, the contractor asks another contractor who is on the tender list for a price which he can quote which will be above that quoted by the interested contractor. Although all selected contractors may not know which of those on the list have taken a cover price, the fact that at least the one asked will know, reduces the effective number of firms. There may be situations, when demand is strong, when most of the prices submitted are cover prices.

It should be pointed out that there is no need to assume collusion in order to have a situation where the contractors divide the market between them. There must, for example, be many situations where, in a relatively small community, if one builder tries to extend his share of the market by reduction of prices and is efficient enough to do so, there will be some form of retaliation from other local builders. This is a classical situation in oligopoly and is illustrated by the kinked demand curve in Figure 12.5. In order to avoid a price war, builders may, in these circumstances, refrain from 'poaching' on each other's traditional clients for fear of similar 'poaching' on their own clients. In such a case, profits are likely to be above normal. The continuance of this situation is, however, restricted by ease of entry.

If a contractor is chosen by the client, either by a simplified tendering situation or by any other method, then there is clearly only one firm in the process of final price determination and, based on the number of firms only, this would lead to a monopoly position. In fact, as in so many cases, the power of this position is limited by freedom of entry, not in this case entry into the market, but by the power of the client at any time either to stop negotiations with the one chosen contractor and invite other contractors to negotiate, or to commence the whole process of selection again by going out to tender. The monopoly power of the negotiating contractor will be greater, the greater the time and money already invested by the client in these negotiations, and this money value is a fair measure of the maximum monopoly profis which this contractor might earn.

Once again, the situation is different for non-traditional methods of procurement. The number of firms capable of organising, say, a design and build project is less than those capable of running a straight construction project. The number able to put together a BOOT or PFI project is even less. Moreover, the clients for the large sophisticated projects, which lend themselves to these methods of procurement, are themselves limiting the number of firms able to compete, by initiating stringent, time-consuming and expensive methods of developing panels of contractors, consultants, material suppliers and so on who are deemed capable of undertaking an efficient operation and are intended to be on the panel for a number of years, after which the list will be reviewed. This is the situation, for example, with various partnering arrangements, frame work agreements, as used by British Airways Authority (BAA), and prime contracting. The reason for these measures is the raising of standards, and it may well facilitate that, but they are also reducing the number of firms in competition.

Many of these schemes in operation in the UK are new and it remains to be seen how they will operate in practice. It may be that the costs to the client of the selection process, together with the costs to the contractor or other party to the process, will limit the frequency of the updating of the list of selected organisations, in which case the effective number of firms in that particular market may dwindle over time. There is, in any case, a danger that the small number of firms, all known to each other, will reduce competition because each firm will perceive that it is operating with a kinked demand curve and will not wish to 'rock the boat' by putting in the lowest possible price. As pointed out above, there is no need for overt collusion for this situation to develop. It could be argued that this situation to some extent offsets the power of the continuing, knowledgeable client by giving some advantage to the contractors and consultants. It does not, however, seem a satisfactory way of doing that. However the situation develops, the new arrangements are likely to change substantially the operation of the market for large projects.

Ease of entry and contestable markets

One of the reasons why construction has been regarded as a competitive industry is that there have been no formal restrictions on entry. Moreover, at the lower end of the industry, it was possible to set up in business with negligible equipment – perhaps a spade, a wheelbarrow, a ladder; working capital is not needed for materials as it is provided largely by builders' merchants' credit; working capital for the builder's own remuneration or equivalent of wages is barely necessary, as the average duration of small jobs may well be about a week and some householders partly finance this by paying for 'materials' in advance.

Once this elementary stage is passed, graduation for the more efficient rather higher up the scale is not difficult, although there may be crisis points in their progress (see Chapter 10 on costs).

As important as, and, in the more specialised and large-scale project markets, probably much more important than, movement up the scale from small to large, is sideways movement from operating in one specialist market to another. Most of the large contractors are in any case working in a considerable number of different markets simultaneously, and if they suspect that in work of the appropriate size the profits are abnormally high in a particular specialism, they are likely to buy in expertise in management and enter the market. Similarly, a small builder from one town, hearing that the market in a town nearby seems to be particularly profitable, may well move in to obtain work there. He will have little to lose compared with the established group within the town. It is suggested therefore that if the situation exists where, because of a small number of firms in a market profits are abnormally high, freedom of entry and relative ease of entry from a separate but similar market will ensure that the low level of competition does not exist for very long.

This suggests that, as a general rule, construction markets are contestable and that ease of entry ensures that high profits due to some degree of market power are unlikely to persist. A study by Ball and Grilli (forthcoming) supports this view. They observe that, if markets are contestable, it would be expected that prices in each market would move in tandem. If contractors in one market were obtaining better prices than other contractors, competitors would move in, and reduce prices to the average level. Ball and Grilli compare pre-tax profits, turnover and net assets from the accounts of 15 medium-size contractors with output of new work, all work and GDP. They conclude that 'price changes in the sub-sectors seem to be determined by changes in the overall demand for construction – rather than changes in demand in the sub-sector itself'. This may now be changing a little. At the bottom end of the market the registration schemes will make it more difficult for firms to move up the scale and, even if they do so, will reduce the speed with which they can react to a change in the market situation. Ball and Grilli deliberately excluded firms with turnover greater than 1 billion or roughly the top ten contractors, in 1994. It is at this top end of the market that the necessity for pre-qualification to get on lists for non-traditional work types, as described in the section above, will inevitably hamper the access of new firms to the specific markets. Thus, markets will cease to be contestable in the short run. Unfortunately, in the construction industry the short-run reaction is important because of the speed with which the level of demand can change. Moreover, it may be that there are no firms capable of entering the market who are not already functioning in it.

Perfect knowledge

Clearly, however efficient the construction industry grapevine, there is not perfect knowledge among the contractors of what is happening in the market. Indeed, the whole system of price determination in construction, ranging from open tender to negotiation, ensures that perfect knowledge of which firms are interested in the job and what prices firms are prepared to quote does not exist. Even after the contract is awarded, particulars of the price quoted by each competitor are not automatically made available by the client, especially in the private sector. It is perhaps understandable that there may be a wish to withhold data on negotiated contracts. Some of the secrecy as to the list of tenderers before tenders are due, helps to prevent collusion and is a case where more perfect knowledge could cause a movement further away from competition. Some other limitations on knowledge, and particularly that on tender prices after the contract has been awarded, increase uncertainty so that work may go to the best assessor of an uncertain situation rather than the most efficient to undertake the work.

Clearly, the market price does not exist in construction in the same way as the concept is applied, say, to a grain market where the conditions are as near to perfect competition as possible. The product of the contracting industry – assembly, transport, management – is much more difficult to quantify and there is no satisfactory unit of measurement to which to apply the price. Even in the ultimate product of the construction industry, namely the building or other construction work, the product is so variable that when a contractor is making a bid he may have no past data on market price. He does not know the ruling market price but is constantly having to guess at it.

Some of the complications of the uncertainty in this and related situations are dealt with in Chapter 14. It is sufficient here to say that on the criterion of perfect knowledge, perfect competition certainly does not exist in construction.

Assessment of extent of competition

An examination of the extent to which the conditions for perfect competition are found in the construction industry has shown that no statement can be made which is true for all markets, for all methods of choosing the contractor and of price determination. Table 13.1 shows, for each type of market, each way of selecting a contractor and each stage in the selection process, the type of market conditions. Where there are many firms in the market the types range from near-perfect competition to some limited monopoly – limited, that is, by the ability of the client at any time to remove the contractor from that semi-monopolistic position in which he has put him. The term 'partial oligopoly' is used to denote that the situation is one of oligopoly, in the sense that the behaviour of the firm at that stage of the process is influenced by its expectations regarding the behaviour of other firms. It is not full oligopoly because each firm in the tender situation does not have such a large share of the market that his output makes a significant contribution to the total. In all cases in the first part of the table in which there are many firms in the market, the extent of the power to make higher than normal profits is tempered firstly by the

Table 13.1 Assessment of type of market in contracting

Type of selection	Stage of selection	Number of firms	Product differentiation	Type of market
		I. MANY FIRMS IN THE MARKET		
Open tendering	Tender	Many	None	Approaching perfect competition
Selective tendering	Pre-tender	Many	Substantial	Monopolistic competition
	Tender	Few	None	Partial oligopoly without product differentiation
Two-stage tendering	Pre-tender	Many	Substantial	Monopolistic competition
	Tender	Few	None	Partial oligopoly without product differentiation
	Negotiation	One	n.a.	Limited monopoly
Negotiation	Pre-selection	Many	Substantial	Monopolistic competition
	Post-selection	One	n.a.	Limited monopoly
		II. FEW FIRMS IN THE MARKET		
Open tendering	Tender	Few	None	Oligopoly without product differentiation
Selective tendering	Pre-tender	Few	Substantial	Oligopoly with product differentiation
	Tender	Few	None	Oligopoly without product differentiation
Two-stage tendering	Pre-tender	Few	Substantial	Oligopoly with product differentiation
	Tender	Few	None	Oligopoly without product differentiation
	Negotiation	One	n.a.	Limited monopoly
Negotiation	Pre-selection	Few	Substantial	Oligopoly with product differentiation
	Post-selection	One	n.a.	Limited monopoly

knowledge that until the contract is actually signed the client can go back to an earlier stage in the process and bring in more competitive firms, and secondly by the fact that the firms want to be selected for other subsequent contracts and therefore must continually sell themselves as good, moderately priced contractors. In the last resort, the client can withdraw his project from the market altogether – at least for a time.

In the second part of the table, dealing with markets in which there are only a few firms, there is oligopoly or, in negotiation, monopoly. It is of little help to the client to revert to an earlier stage in the tender process because, even though entry is easy, it may take time to attract firms from other markets. However, the ease of entry is probably of sufficient importance to prevent firms charging prices significantly above normal, for any length of time. In this situation too, the client can withdraw his project, and may well do so if the price seems to be too high.

The diversity of the other procurement methods makes it difficult to include them in the table. Most of them would be in the category of few firms with considerable product differentiation and would therefore be classified as some form of oligopoly. It seems that many of the schemes for procurement, which include finance and the operation of the installation provided, could lend themselves to oligopolistic behaviour, as illustrated by the kinked demand curve (see Chapter 12). It might be, in practice, that the suppliers never get the opportunity to make above-normal profits because their power is balanced by the monopsonistic power of a knowledgeable client. It would, in any case, be almost impossible to judge whether profit was above normal levels because of the very high level of risk involved in the provision of a service over a long period of years, especially one of great complexity and difficulty of definition, such as providing and running a hospital. A further problem is that the costs of tendering for such projects for all the parties involved are extremely high (see Chapter 14). The costs of the unsuccessful tenders have to be borne by the successful ones.

Level of profit in the industry

One measure of the degree of monopoly power is the extent to which the profits in the industry are higher than normal. A study (Hillebrandt, 1984) of the return on capital employed and on turnover of large companies in construction and other industries generally show a rather lower than average profit on turnover, and probably the differences in profit on capital are not significant. The low capital employed in relation to turnover and the high risks of construction affect the situation. Akintoye and Skitmore (1991) find that the level of profit on turnover of 80 construction firms studied between 1980 and 1987 was about 3.2 per cent. They comment, 'Compared with other industries, this may be considered to be rather low, especially as the 3.2 per cent is a pre-tax profit margin.' The margin of larger contractors was higher, partly owing to their diversification into other businesses, notably house building, which was more profitable than contracting. The authors regard this figure of about 3 per cent as having always been the level of profitability in construction, quoting studies in the UK and the US including Flanagan (1990) showing that profitability in the construction industry seems to have been hovering around 3 per cent irrespective of the state of the market. This conclusion may be invalidated by the very low profits made in the early 1990s. According to Hillebrandt *et al.* (1995), pre-tax profits in contracting of 70 of the top 80 UK contracting companies in 1993 were about 10 per cent of those in 1990. These last data support the scepticism of Ball and Grilli (forthcoming) about building contracting producing long-term constant rates of return. However, it is possible that that low level will not be repeated and that the industry will return to the long-run average rate of return.

There is no evidence that large firms in the construction industry are making exceptional profits. If there are situations where there is monopoly power in the industry, they must be relatively unimportant compared with the construction industry as a whole.

Influence of the client

The influence of the client has been discussed above and in Chapter 8. It is clear that actions of the client can substantially determine the degree of competition. There is a great variety of client in the construction industry, from the small unsophisticated client building for the only time in his life, to the large client building more or less continuously and with sophisticated professional advice, often his own employees, at his disposal. The first type of client is likely to be numerous, leading to conditions approaching perfect competition among buyers – although they would certainly not have perfect knowledge. The second are likely to be few, with, in some markets, monopsony; an example is the government, as a purchaser of defence installations.

In view of the fact, however, that almost all construction projects are separately let, and therefore for a given project the client is the monopsonist, the significance of the nature of the client lies in his degree of sophistication and market knowledge, so that if conditions approaching perfect competition do not exist in that market, he can use his knowledge and power to obtain favourable prices from the existing group of suppliers and, if necessary, increase the degree of competition by going outside the existing suppliers to, say, for a large project, international competition or, for a smaller job, one of the national contractors. If the client is a householder in the repair and maintenance field, he may in a sense be a monopsonist but he probably has not the knowledge to use his power.

Slope of demand curve facing the contracting firm

As explained in Chapter 12, under conditions of perfect competition the average and marginal revenue curves of any firm are horizontal and identical and are equal to the price in the market. In the case of monopoly, and to some extent in all types of market situation other than perfect or pure competition, the average revenue curve slopes downwards, which means that as output is increased, the average revenue or price received per unit of output decreases. The slope of the average revenue curve for firms in the construction industry must now be considered specifically.

Effect of the type of market

The position in the contracting industry is complicated particularly by the fact that, as shown in Table 13.1, there is, except in open tendering, some non-price competition in which virtually all the firms interested in that particular market at a given time participate and that, once the client has made his selection of one or a group of contractors, price determination takes place among this limited group only. In the case of markets in which there are a few firms only, the number of firms engaged in the price-determination process may be all or nearly all those firms in that market, and therefore in this situation the two processes merge. The only difference is that in the pre-tender stage there is advantage in stressing product differentiation, to ensure that the firm is chosen for the price-determination stage. This will be unimportant if the firms are so few that the client has little choice but to include all.

In the latter situation, therefore, there is a fairly clear position of oligopoly. The demand curve of each firm will slope downwards to the right because its share of the market is significant and therefore related to the industry demand curve. It was found in Chapters 4–8 that the industry demand curves for construction tend to be fairly inelastic. The demand curve will therefore slope fairly steeply downwards (as D, in Figure 4.3). Thus, the slope of the demand curve or average revenue curve of the firm will have a significant downward slope – although less than that of the industry.

In the case of non-traditional arrangements, the product differentiation continues through the whole process. Furthermore, there are often very few firms in the market so that the slope of the average revenue curve may go steeply down. This means that a firm can do better by providing a lower output at a higher price than if the demand curve was nearer to the horizontal, but it can only do this if its actions do not start a competition war with its relatively few competitors. This suggests his demand curve is kinked.

In the former cases of many firms in the market, the extent to which the demand curve slopes downwards is dependent on the degree to which the firms in the market succeed in differentiating their product so that, when the firms are chosen for, say, selective tendering, the client sees this group as producing a product significantly different from those of the non-selected group. If he does, then the firms have succeeded in carving off a little part of the total market and in forming a limited monopoly or oligopoly in it. In this case this part of the market becomes in effect a new market on its own and the demand curves of the individual firms will be related to the demand curves in the market. However, as other contractors have products which are fairly close substitutes for theirs, the slope of the new minimarket demand curve will be rather elastic and therefore the slope of the derived demand curves of the individual firms within the market will slope only very slightly – perhaps insignificantly – downwards on account of the market situation. This means that in order to get to the stage of persuading more clients to buy his particular variety of contracting, the contractor will have not only to persuade them of its merits but also to make its price attractive as well. Similarly, if his price is rather high, he will lose a large number of potential clients. It is important to realise that this particular price persuasion operates mainly when the client is considering which contractor to short-list. In the actual price-determination situation the position is rather different.

In the case of non-traditional procurement measures, the market has to some extent been reduced in size by various longer-term arrangements, for example partnering or some arrangements, still to be determined, for prime contracting. In any case, the greater the expertise required of the contractor, the smaller the market becomes. When the stage of PFI is reached with contractors expected, in some way or another, to be able to cope with finance and operation of the facility as well as the more familiar design and construction, the number of contractors in that particular market is substantially reduced.

Tendering data

The practical extent of competition may be looked at in a different way. If a firm is at a given point on its average revenue curve, and had been, say, putting in tenders with a mark-up of 3 per cent and had obtained 100 units of business, the entrepreneur could probably give an answer to the question: If you had put in a mark-up of 1 per cent, what volume of business would you have obtained, and what would it have been with a 5 per cent mark-up? At the lower price the firm would probably have obtained more business, and at the higher price less business.

A study of tenders for office projects in the UK in 1982–3 by Flanagan and Norman (1989) found that on contracts over £1 million, more than 80 per cent were within 10 per cent of the lowest bid, compared to less than 60 per cent for contracts under £1 million. This means that a small change in the mark-up would have a very substantial effect on the number of contracts obtained. The very fact that the contractor can obtain more work at a lower price than at a higher price, invalidates any claim that construction is near to a perfectly competitive market.

Possible reasons for a wide range of tender prices for any contract are, firstly, that the estimators of the various firms do not agree on the likely cost of the project. This is discussed in detail in Chapter 14 but the range of cost estimates is frequently substantial. Secondly, the contractors may have different levels of cost, implying that some are substantially more efficient, and therefore have lower costs, than others. This is possible, but difficult to assess in view of the fact that they cannot estimate their costs accurately and the difference in costs is probably small compared with the differences due purely to wrong estimates. Thirdly, the competitors may be at different points on their cost curves. Fourthly, the contractors may have added widely different mark-ups for overheads and profits. Assuming that they would like to obtain the contract, this implies that they have very diverse estimates of the market price. It should be noted that the first and fourth of these reasons are incompatible with assumptions of perfect knowledge in perfect competition, and that all imply substantial differences in the cost curves of the firm and/or of entrepreneurial ability.

Non-resaleable product and single-project market

The contractor is in a different position from manufacturing industry in that he sells his product, namely the service of assembly, and so on, to a particular client, but that once that service is sold, it ceases by its very nature to exist in that form and becomes embodied in the building (or other work) which is created. This building is custom-built; that is, it is built at a specific place and in a particular way for the client for whom it is intended. The client can sell this building to someone else, but (leaving aside the effects of inflation) it would normally be expected to be less valuable to the other person, because it was not designed especially for him. The client cannot resell the services of the contractor because they have been used up. Thus, as in the case of other services such as doctors' advice, there is no market in which resale can take place. This means that the contractor has some of the advantages of a discriminating monopolist. In a limited sense, each price-determining situation, i.e. the tender or the negotiation, for a single building is a temporary market and the contractor can get as high a price as possible in this market with no risk that the client with whom the bargain is struck can resell the same product, i.e. the service, in another market. But in this one-project market there is no average revenue curve and no marginal revenue curve, for there is no possibility of varying the units of output which will be sold. It is an all-or-nothing situation. Moreover, the price fixed in this one-project market does not affect the price of previously struck bargains in other markets. It is the contractor's interest to make the price as high as possible while still obtaining the contract. This is dealt with in some detail in Chapter 14. In more general terms it is elaborated in Chapter 15, where revenue and cost are considered together to determine how a contractor may best attain his objectives.

14
Price Determination for a Single Project

In the construction industry, price is determined for large indivisible amounts of work, each one of which may represent a large proportion of the work load of the contractor or of that part of his organisation operating in a particular market. The most usual form of price determination is some form of competitive tendering, but negotiation is important and the whole process is complicated by the wide range of non-traditional processes (see Chapter 8).

The contractor may be quoting a price not only for the construction of a building or works, but possibly also for the design, financing and management of the building and/or of the activities which take place in it. This, in turn, may make it difficult for the client to make a sound judgement and consequently the tendering or negotiation process is often carried out in various stages, as mentioned later in this chapter. The client may make his choice on a number of factors, of which price is only one. Whilst it is probable that for most contracts the tender with the lowest price is accepted, this is not by any means always the case. Indeed, in some countries, for example, Italy, the practice is not to adopt the lowest tender but sometimes the average, the second lowest, or the choice is made on some other basis (DL&E, 2000). In some countries, such as the USA, contractors are encouraged to submit alternative design proposals. Savings in costs might later be shared (DL&E, 2000).

On most projects, contractors are assessing their costs and their price before the work is done at the prices of the time. If the contract is a fixed price contract, the contractor must forecast the expected change in prices over the life of the contract. The whole process is, in fact, the reverse of the procedure in most of manufacturing industry where the producer has finally to determine his price only after the good has been produced and his costs are known. Moreover, the manufacturer normally controls the conditions under which his product is sold, whereas, in construction, the client is the main party fixing the conditions of price determination and the later contractual arrangements. This has arisen partly because the client initiates the process and partly because the client is, for the particular product under consideration, the only buyer or a monopsonist. He has considerable power throughout the process, though if he finds no contractor willing to tender on his conditions he will have to make adjustments. The contractor has to put in a bid for each project against the realities of his business situation, including how near he is to his optimum level of output as determined by his cost and revenue curves. This background situation will determine how important it is for him to have the work and, hence, at what potential profit he is prepared to tender.

There is a considerable range of the value of bids for any project and there are many possible reasons for this. Firstly, there would be a considerable range of cost estimates, even if the estimators were working for identical firms, due to errors and different assumptions about, for example the future trend of prices. Secondly, the firms may be at different points on their cost curves so that their costs are genuinely different. Thirdly, the contractors may have very varied views on the state of the market, and therefore on the mark-up which should be applied to the cost.

Once the contractor has been given the opportunity to compete for a project, there are four stages in the process of a contractor's response, although they may in practice run into one another. These are:

- the decision to bid or not to bid;
- determination of the cost of the project;
- assessment of the lowest worthwhile price (or lowest mark-up) at which the work should be taken;
- determination of the final mark-up.

Each of these will be considered in turn.

To bid or not to bid

Before deciding on the cost and price of a project, the contractor must decide whether he wants the job at all. This will partly depend on his work load in relation to his optimum level of output. It might be that obtaining that contract would take him into output levels where his costs are rising rapidly. If there is a danger that the price he could realistically expect for the project will not cover his costs, he should not tender at all. In any case, he should take into account the costs of estimating which, although only a small percentage of the contract value, could tip the balance against putting in a bid.

A study by Shash (1993) of 85 top UK contractors found that the most important factors for the decision to bid or not to bid are the need for work, the number of competitors tendering and experience in such projects. Other factors considered important include project type and size, owner/promoter/client identity, contract conditions, type of contract, past profit in similar projects, tendering method, risk involved owing to the nature of the work and availability of qualified staff.

The cost of the project

The estimating process requires a very detailed analysis of cost. The traditional way in the UK has been based on a bill of quantities; that is, an analysis of all the items in the project prepared by the quantity surveyor in the client's team. The contractor then prices each item before he puts in his tender. The bill of quantities may be used as a basis for tendering, measuring work done and assessing final accounts, including claims. The procedure is different from that in a number of other countries where tenders are based on drawings and specifications or on a simpler bill. In the UK, the bill of quantities is used less than formerly. A survey undertaken by Davis Langdon & Everest on behalf of the RICS (DL&E, 1999) found that in 1998 only 32 per cent of the contracts studied included a bill of quantities.

Whatever the process, it is extremely difficult to assess accurately the cost of every single item in a large project and it is virtually certain that there will be mistakes and errors of judgement. For smaller works undertaken by smaller firms, the estimating process may be very broad brush. In addition, it is quite possible that one contractor may be able to undertake a job at a lower cost than his competitors because he is at a lower point on his cost curve. One of the difficulties of the tendering process is that the lowest tenderer is often the one that made errors of judgement, probably on the assessment of risk, rather than the one genuinely able to do the job at the lowest cost and mark-up. That is why some countries adopt various other tendering systems in which the job is won by, for example, the price nearest to the average or the second lowest price. However, where the system is to accept the lowest tender, contractors always have to consider that they may have made a mistake. Their only escape from this situation, if it is realised in time and if the applicable law allows, is to withdraw the tender. They may also put in their estimate an 'error protection mark-up' separate from and additional to their profit mark-up (Gruneberg and Ive, 2000).

There are also other considerable risks and uncertainties in estimating cost, however it is done. In this context, risk arises when the assessment of the probability of the occurrence of an event can be calculated. Risk is insurable. Uncertainty arises when the probability of the occurrence or non-occurrence of an event is indeterminate. Uncertainty is not insurable (Knight, 1921). Examples of risk are the occurrence of high rainfall in the month of February or of a fire breaking out on site. Uncertainties include the possibility of a manager not performing well, a subcontractor becoming insolvent or an unexpected rise in material prices, for example, because of a rise in the price of oil. Such items would not be included in the estimators costs but would be borne in mind in considering the overall mark-up.

There is another factor to be considered, namely that the final price is not necessarily the same as the tender price because, for example, of variations from the client, possible liquidated damages from the contractor or unspent contingency allowances. Contractors normally take this into account when tendering, sometimes even gambling on the likelihood of variations. However, in England on local authority contracts in 1996, it was found that one quarter of projects cost more than five per cent above the contract sum and one quarter more than five per cent below (Audit Commission, 1997).

Bearing all these factors in mind, it is not surprising that the assessment of cost made by estimators of the firms in competition are often very different. It can be that estimators are influenced by recent high rates of failure to obtain work for which they have tendered and lower their cost assessment, perhaps subconsciously, to compensate for their run of bad luck. In large firms the estimates would be checked and such bias avoided, but in small firms there is no one to do the checking. In any case, the actual final cost of a project is affected by a number of diverse characteristics, including the size, complexity and duration of the project, compared with the usual range of projects of the firm. These factors do not directly concern the client, except possibly the duration of the job and hence its intensity. Consider the situation in which the client wishes to occupy the building at a particular date – say a school to be available in August for September. If he has the building earlier he will incur interest charges on the capital for the whole project when he expected to build up to this figure only in August. These additional costs are shown in Figure 14.1 as linear, i.e. *CD*. In fact, taking into account the discounted value over time they would not be linear, but this problem has been ignored in this diagram for the sake of simplicity. If the client has the building later than August he will lose all the advantages of September occupation and the opportunity cost of not having the building will rise sharply as *DE*. On the other hand, in the situation in which the client wishes to use the building as soon as possible and will obtain rent for it (or save rent elsewhere), the rise is opportunity cost will be linear as *FG*.

The contractor's costs are, as already explained, as *AB* in Figure 14.1 with optimum completion time *P*. The client will have to bear the contractor's costs as well as a higher mark-up, so he wishes to minimise the value of *AB* + *CE* or *HJ* in the first situation and the optimum completion time will then be at *K*. In the second situation he wishes to minimise the value of *AB* + *FG* or *LM* and the optimum completion time will be at *N*.

Under the normal tendering system neither the contractor nor the client knows the time/cost curves of the other and the chances of arriving at the optimum rate of construction are slight. In negotiation there is less of a problem because the time of delivery would be one of the matters for discussion.

Lowest worthwhile bid price

After considering all the risk and uncertainty in the determination in the cost of a project, the contractor has probably in mind a range of possibilities around his best cost estimate of what the final cost might be on the worst case and best case scenarios. From this viewpoint he must decide what is the lowest price at which he would be prepared to do the job, and therefore what is the lowest price at which he would bid. This will depend on many of the factors already considered in the decision as to whether to bid or

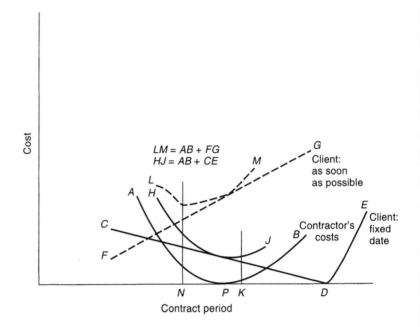

Figure 14.1 Additional costs associated with time

not. Shash (1993) found that the degree of difficulty, the risks involved and current work load were put at the top of the list of
factors to be considered in the mark-up. They would also be the
most important in the decision on the lowest worthwhile bid price.
The personality of the contractor and his willingness to take risks
will also affect the situation, as will broader economic factors, such
as expectations about future market trends.

The final mark-up

Having decided that he would like to obtain the contract at a certain minimum price, the entrepreneur must then balance the mark-ups
he could add on against the likelihood of obtaining the job at these
prices. The contractor may still be worried about some of the factors he has considered earlier in deciding to bid and deciding the
lowest worthwhile bid price, so he might prefer a higher mark-up
if it does not reduce too much his chances of getting the job. While
this particular contractor has been considering his position in rela-

tion to this tender, his competitors have, in all probability, been doing exactly the same. The chance of this one contractor getting the job will therefore depend on the state of the order books of his competitors and where they stand in relation to their optimum level of output, how they view the conditions of the tender, how good they are at estimating costs, to what extent they are risk averse and their view of the future levels of demand. Most contractors have a view of the likely market price of a project and may even fill in the bill of quantities to make the total fit their assessment of the market price. The assessment by competitors of the likely market price may be different and this too can lead to very different final prices. Clearly, it is important to know as much as possible about one's competitors. The chance of being successful in a tender will also depend on the number of competitors because, statistically, the more there are, the less chance any one has of being successful.

Bidding theory

In the late 1950s and early 1960s a body of theory on bidding was developed, mainly in the USA. Basically, these theories attempt to determine what mark-up contractors should add to their estimated cost to maximise the profit in the long run and assuming that they continue bidding on a consistent basis. They incorporate available data on the history of bids against known individual competitors or unknown competitors where the range of past bids is known. Calculations are done by multiplying various mark-ups by the probability of getting the job at that price.

The work was originally based on papers by Friedman (1956), followed by Gates (1967) with an alternative model. There followed a period of intense activity in developing variants of these models, some based on profit maximisation and others, for example Broemser (1968), with utility maximisation. Park (1979) and Park and Chapin (1992), using Friedman's model, present a very clear description of the method. Arguments between the protagonists of Friedman's model and Gates' model continued into the 1980s. Gates (1983) came to the conclusion that these models, although of interest to academics, were not helpful for the practitioners and moved towards a model based on the formal assessments of subjective judgements of several experts, known as Expert Subjective Pragmatic Estimates (ESPE).

Unfortunately, the methods of these earlier models are too simplistic. Taken on their own, the methods ignore many factors taken into account by contractors in considering their bid price. A helpful review of the current situation in tendering theory by Runeson and Skitmore (1999) highlights some of the problems. The most serious is that it assumes constant mark-ups and that these are unaffected by variations in demand. Many of the factors considered by contractors, and described above, relate to changes in demand, including their current order books which will have been affected by demand and even expectations on the future market situation. Moreover, the history of the last decade in the construction industry in the UK paints a totally different picture. During the recession of the early 1990s, or more precisely from the third quarter of 1989 to the third quarter of 1992, tender prices of construction in Great Britain fell by 23 per cent (BCIS, 1999), in spite of there being no apparent fall in the cost of materials or labour (DETR, 1998b, Table A). How can that have happened if contractors were not putting in lower tender prices in the hope of winning more contracts? Any contractor who continued putting in the same mark-up as before would have had no work and been quickly out of business. Clearly, there is a market in construction services and the level of prices in that market is determined by demand and supply. The only mechanism for actually setting those prices is the tendering, or other price determination methods, for the individual contract.

Some analysts, for example Grinyer and Whittaker (1974), have tried to overcome the problem of changes in the market situation by asking the entrepreneur to estimate a market coefficient; that is, the arithmetic mean of what competitors' bids will be. This is, in fact, reinstating individual judgement as an important factor but is, at the same time, partly dispensing with the need for the model.

Runeson and Skitmore conclude that bidding theory, as it exists at present, is not soundly based. Moreover, Shash (1993) found that the number of competitors and their competitiveness were well down the list of the factors considered in determining mark-ups, although they are the mainstays of probability bidding theory. It is perhaps not surprising that, according to Shash, mathematical or statistical models were used in 1990 by less than 18 per cent of 85 contractors in his sample of large and medium UK firms. The others rely on 'experience, judgement and perception'. In 1988 less than 11 per cent of top US contractors used the mathematical models (Ahmad and Minkarah, 1988).

Alternative approaches to bidding decisions

An alternative approach was developed by the author in the first edition of this book. It is based on ideas and concepts developed by Shackle in a number of publications (for example, 1952, 1955) and was adapted and extended to apply to the specific construction industry tendering situation. The method develops a theory of the stages of thought through which contractors go in order to arrive at a decision on tender prices. The theory is another way of thinking about the three decision stages. It does not tell contractors precisely how they should assess their surprise at the occurrence of various events. The analysis is presented in Appendix B.

Benjamin (1969) also looked at the whole process, but in terms of utility. He notes three elements in the competitive bidding problem which more or less correspond to the three stages of the degree of potential surprise analysis:

- a probability distribution to express the uncertainty of the cost of performing the work;
- a non-linear utility function which scales the bidder's preferences for different amounts of money when the possibility of a loss is considered explicitly in the problem;
- a means of assessing the probability of winning the contract with different bid amounts.

Unfortunately, Benjamin acknowledges that the data are not available to feed practically relevant information into his model.

Negotiation

The analysis of price determination in this chapter has so far been presented entirely in terms of the tendering situation. Much of it, however, is equally relevant to other methods of pricing the product of the industry. The determination of cost and of the lowest worthwhile price at which the contractor would wish to undertake the job apply equally to tendering and to negotiation.

The determination of the final price is different. Before negotiation the contractor will have decided the lowest price he will accept, and during negotiation he will attempt to obtain the highest possible price while still getting the contract. The client will have in mind the highest price he is prepared to pay for the job and will have the objective of obtaining it for the lowest possible price while still maintaining the quality of the finished product. As indicated earlier in Chapter 13, in negotiation the contractor may be in a limited monopoly situation, although the client can negotiate with more than one contractor. The monopoly is limited by the knowledge that if the client does not like the way the negotiations are proceeding, either with regard to price or for any other reason, he can go to other firms in the market. Within this framework negotiation will take place and the final outcome will depend on how much each negotiator wishes to deal with the other, on the top and bottom limit to the range within which a deal can take place, and on the skill of the negotiators.

Marsh (1973) analysed the process in considerable detail. He stresses that the final outcome often depends as much on the individual motivation of the negotiators for the two sides as it does on objective or economic considerations. He deals with the question of time costs at some considerable length because they have an important bearing on the outcome of the negotiations. This supports the conclusion reached in Chapter 13 of the importance of the time and money already invested in negotiations on the level of supernormal profit which the contractor may be able to obtain.

Non-traditional methods of provision of buildings or other facilities

Apart from negotiated contracts, discussed above, there is now a range of different methods of organising the process outlined in Chapter 8. These complicate the process of price determination because the scope of the arrangement between the client and the contractor is much greater, including some or all of the design, construction and management of the completed installation. The exact nature of the contractor's obligation cannot be predetermined. As a result, a whole mass of different methods are being used to come to a contractual arrangement. These include selective tendering based on preliminary data and a fee. The fee may be broadly related to the cost of the operation or combined with sharing of savings from a target cost. Alternatively, there may be negotiation with one or more contractors. The contractors' remuneration must ultimately be included in this negotiation.

Provision of a service to include the finance for a project

BOOT and PFI are very different from the traditional construction process because the contractor, who may be part of a bidding group, is selling a service, the price of which is 'calculated as a service charge levied at regular intervals throughout the lifetime of the contract' (CIC, 1998, p. 44). The organisation of the process, as well as the price determination, is extremely complex. For PFI, which deals with public sector projects, the determination of price involves assessment of costs which have nothing to do with the actual construction, which is the area with which the contractor is familiar. In addition to assessment of design and construction costs and their associated risks, the contractor must consider and determine the sources and costs of finance, and the cost of operating the constructed facility. In all assessments the effect of inflation must be considered, as well as possible changes over time in the demand for the service being provided and the optimum method of supplying it. The risk that the contractor will judge wrongly is large and the consequences of being wrong substantial. Mugume (1999) lists some of the risks associated with this type of project and some possible measures of alleviation.

At the outset, the relations between the various parties to the bid must be determined and formalised. The design, for example, may be undertaken by the designer on a fee basis or on a no-win, no-fee basis. If the latter option is taken, the design fee for a successful project will be higher. However, the CIC points out that the latter limits the freedom of the ultimate concessionaire, for example, to commission a design and build contract (CIC, 1998, p. 38). The bidding team will need legal and financial advice on contractual arrangements, and to negotiate with and assess funders to determine the most beneficial financial package. The costs of tendering are therefore very high, even at an early stage, and the chances of success may be small if there are several bidding groups involved. It may be that the contractor, having done the preliminary work, will decide not to bid, because of doubts as to whether 'the project would be at all achievable using PFI, attractive in commercial terms and offer the opportunity for the bidding firm to be competitive' (CIC, 1998, p. 40).

If the bidding team decides to go ahead, there will be a number of meetings with the client, further submissions will be called for and after long negotiations the client will select a few preferred bidders. As the negotiations continue, one or two teams will be asked to elaborate their proposals to the stage of making a definite offer. Even then, the whole process of moving towards a PFI project may break down. The tendering costs of the whole bidding process are substantial. The Construction Industry Council warns that 'the total costs of bidding for the preferred bidder have sometimes been upwards of 10 per cent of the capital costs of the project for smaller projects, with larger health projects costing £3–4 million through to due diligence' (that is, the final enquiry stage by the financiers) (CIC, 1998, p. 35).

Game theory

The situations in negotiation of contracts and in the constant interaction of clients and contractors, even under conditions in which several contractors are in competition, as described above, may be more appropriately considered from a theoretical point of view in terms of game theory than as an extension of any sort of tendering theory. Runeson and Skitmore (1999) refer to Shubik (1955), who argues that game theory applies when the outcome depends not only on their own action and chance, but on the actions of the opponents. Perhaps, when the PFI procurement process has settled down, a game theory tailored to the specific conditions of PFI could be developed. However, this is outside the scope of this book.

15
Conclusions on Costs, Revenue and the Equilibrium of the Contracting Firm

Having completed a study of the costs of the firm and the industry (Chapters 10, 11 and 14) and of the demand facing the firm (Chapters 12–14) and the industry (Chapter 4–8), it is appropriate first to summarise the main findings and then to combine them, and to draw some conclusions on the equilibrium position of the firm on the two main assumptions of the objectives of the firm, namely profit maximisation and maximisation of turnover with a profit constraint (Chapter 9).

Summary of findings on costs

In the short run, the average variable cost curve of the contracting firm will be U-shaped, because at very low outputs there are some difficulties in the efficient use of inputs, and at high outputs in the short run, in which the resources of head office and some site management are fixed, the managerial inefficiencies and organisational problems are likely to cause rising costs of the other inputs, particularly of manpower whose efficiency is especially dependent on good management.

Fixed costs are relatively small in traditional construction, but nevertheless they suffice to make very small outputs expensive, and hence the total average cost curves rise more steeply to the left of their lowest point and slightly less steeply to the right of it than was the case when only variable costs were considered.

There was some discussion of the postponable costs, namely part of the normal return on the entrepreneur's labours, and some of the return on capital invested, depending on whether interest *must* be paid regularly, whether capital *can* be withdrawn from the business and whether the firm wishes to expand.

In the long run, fixed costs can be altered and postponable costs must be met, but there is no reason why the long run should be a particular definite period of time or, indeed, the same for all factors. In the very long run, technologies can be altered and new technologies developed, and in any case when there is a change in the level of fixed costs there will usually be a change in the technology used. The shape of the long-run cost curve is uncertain. There are theoretical reasons, particularly the shortage of entrepreneurial ability and the possible inelasticity of factors of production, why the curve should eventually turn upwards, but instances of large and growing firms show that at least for some it is fairly flat over long ranges of output. Over time it may actually be sloping downwards because of improved technologies. In any case, because of the indivisibility of certain factors of production there are almost certainly humps in the long-run cost curve.

It seems fairly certain that there is a limit to the *rate* of growth of the firm, which means that as output increases or decreases over time the short-run curve is likely to be U-shaped. Moreover, the short-run cost curves are not reversible, in that the cost of increasing the level of output may be different from the costs of decreasing it.

In a consideration of the costs of the single project, the total number of which, over the whole period of their duration, determines the place of the firm on its average cost curve, it was found how extremely uncertain the real costs of a project are. Because of this, and because the contractor does not know in advance whether a certain contract will be obtained, there is uncertainty as to the actual level of his cost curves, and of the point on them where he will be at a given time in the future.

Summary of findings on revenue

The demand curves facing the contracting firm are determined by the demand curves facing the industry as a whole in all the various markets and the extent of competition in these markets. It was found in the chapters on demands on the industry that the demand curves for all types of work were relatively inelastic. Thus, if there were very few firms supplying these markets, the slope of their average revenue curves would be great and their monopoly power would be considerable.

In practice this is not the case, even in markets where there are relatively few contractors. At stages of the process of selection of contractors and of price determination, where one or a few contractors are involved, the extent of the power of the contractor to act as a monopolist is limited by the ease with which other firms can enter that market and with which the client can increase the number of contractors involved. Overall, there is little evidence of monopoly power in the industry, and such situations in which it exists account for a tiny proportion of the work of the industry as a whole. Thus, the demand curve actually facing the contractor is likely, on these arguments, to be fairly elastic.

This is supported by a consideration of the tendering situation where, if a contractor consistently puts in tender prices with a low mark-up, he will get more business than if he puts in a high mark-up. Thus his demand curve cannot be horizontal as under perfect competition. However, if the contractor is not tendering for all the work in the market in which he operates, he would be better off (in spite of the small increase in his cost curves due to higher estimating costs) by tendering for more contracts at the same price than by lowering his price on the same number of tenders. If, therefore, he is able in some circumstances to obtain more work at the same price, his demand curve must be near to the horizontal perfect competition model.

If, however, the contractor is already tendering for all possible work within a given market, then, subject to the safeguard of ease of entry to any market, the average revenue curve may slope rather more steeply. Another possibility is that the contractor in such a market may have a kinked demand curve: that if he lowers his price the other contractors will take this into consideration in their assessment of the chances of obtaining work in the tendering situation and lower their price still further. If he raises his price, however, there will be no particular reaction on the part of his competitors.

Synthesis of costs and revenue: short run

Combining the likely cost curves of the firm with its probable revenue curve it is possible, following the analysis of Chapter 12, to see the way in which the firm maximises its profits. In Figure 15.1 are shown the average revenue curve. *AR*, the marginal revenue curve *MR*, the average cost curve *AC* and marginal cost curve *MC* of the firm. Assume that the cost curve relates to the period of time in which it includes normal return on the entrepreneur's labours plus normal return on capital which can be withdrawn from the business but in which the fixed factors of management and head office cannot be altered. Then the output which maximises profit in this period is that at which marginal revenue equals marginal cost, that is, at *A* in Figure 15.1A, at which average revenue and hence price is 210 and output about 4.6 units.

Maximum output making only normal profit is that at which average cost, including normal profit, equals average revenue, or point *B* in Figure 15.1A, at price 125 and output about 8 units.

In Figure 15.1B is shown the same situation in terms of total costs and total revenue. This presentation is more helpful when some additional level of profit is looked for, other than the normal profit included in the cost curves. The difference between the total revenue curve and the total cost curve is profit (in excess of normal profit) and is plotted as *TP*. Thus, the output which maximises profit is about 4.6 units, as was found from Figure 15.1A, and that where there is no profit above normal profit is 8 units. If the entrepreneur wishes to obtain some level of profit below the maximum, in order, for example, to give more benefits to his employees, the output which he needs to obtain this level of profit is clearly ascertainable. Thus, if he wishes to obtain a profit of, say, 200 above that included in the cost, combined with as high a level of turnover as possible, then he should be at *C* where his output will be about 7.3 units.

It was seen in Chapter 14 that the mark-up appropriate to any particular tender depends on a number of factors, including the state of the firm's order books, and where they expect to be on their cost curves relating to the period over which the work on the contract will extend. A study of the relationship of the marginal cost and marginal revenue curves in Figure 15.1A shows how important this assessment can be. The difference between the marginal cost and marginal revenue at a point in time (say, for practical purposes a month) varies greatly according to the work load, and therefore the mark-up which the contractor should contemplate to obtain that work also varies considerably. The situation is complicated by the fact that, because of the irregular work load over time (see Figure 10.2), the place on the cost curve will vary from month to month, and yet a decision has to be taken for the contract as a

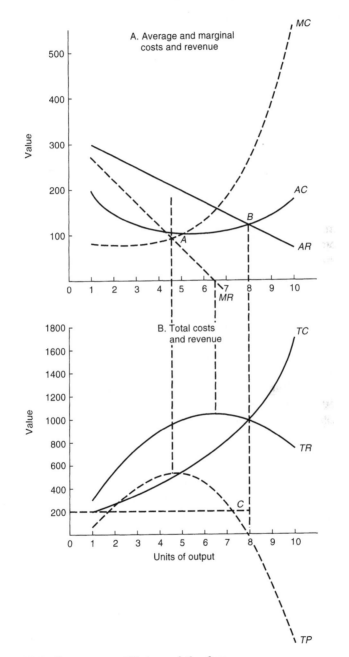

Figure 15.1 Short-run equilibrium of the firm

whole. It is, however, clear that to adopt a policy of a constant mark-up is justified in only very special circumstances.

Synthesis of costs and revenue: long run

In the long run, the average cost curve of the contracting firm may, as suggested earlier, have a very stable overall level but with 'humps'. If this is so, its situation would be rather as that shown in Figure 15.2A, with AR_1 and MR_1 representing the average revenue and marginal revenue curves as before in Figure 15.1A and AC representing the humped average cost curve and MC the corresponding marginal curve. In this diagram the marginal revenue curve MR_1 cuts the marginal cost curve at A, at which the appropriate output is 4.8 units at which profits are maximised. In terms of total costs, the situation is shown in Figure 15.2B. TR_1 is the total revenue curve corresponding to the average and marginal revenue curves in Figure 15.2A. TP_1 is the total profit curve derived from TR_1–TC, the total cost curve. Thus, if the entrepreneur wanted to maximise profits he should not employ the new fixed factors necessary to expand output. If he wished to maximise revenue, subject to making a normal profit, he should move to point B with output 8.5 units, i.e. he should go over the 'hump'. If, however, he wished to make a minimum profit of 200 and have an output as large as possible subject to this constraint, he should expand his business over the hump but produce only 6 units of output.

The illustrations above are all in terms of the total revenue curve TR_1, which is derived from the average revenue curve in Figure 15.2A. If, however, the market is small and the entrepreneur has a large share of it, it is possible that he anticipates an upward shift in the demand curve so that it might become profitable for him to expand output to be in a position to take advantage of this. Thus a contractor in a developing country producing housing might expect an expansion of demand so that the total revenue curve would shift upwards and to the right, thus making it more profitable to produce at larger outputs beyond the 'hump'.

In markets where the firm has a relatively small share, the total revenue curve may not turn down but over the relevant range of output continue to slope upwards at a rate faster than the total cost curve, as TR_2 in Figure 15.2B with a total profit TP_2. There would be no optimum position within this range of output. This may be the state of some of the large contractors who have over-

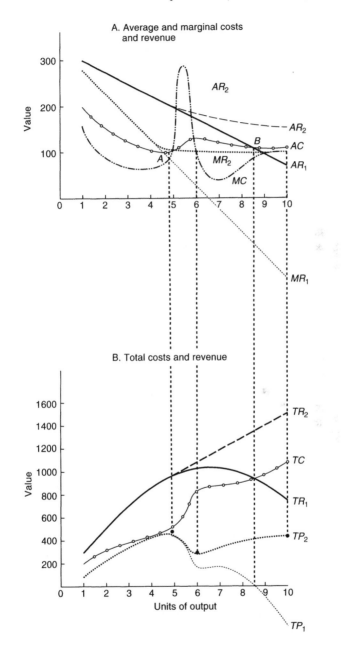

Figure 15.2 Long-run equilibrium of the firm

come most of the 'humps' and who could, in any market which was large in relation to their output, go on expanding both output and profit. As was seen under the heading 'Summary of Findings on Costs' above, there are reasons why in any market this state of affairs is unlikely to continue indefinitely.

Most contracting firms operate in several markets. The overheads of the business as a whole, which cannot be allocated in any sensible way, should be arbitrarily apportioned to the business in each market as fixed costs. In this way, they will not affect the marginal cost. In order to maximise profit, the firm should, in each market, operate at the output which equates marginal revenue and marginal cost. If the demand curves facing the individual firm are different in different markets, or if their cost curves are different, this may result in charging a higher price in some markets than in others. As was seen in Chapter 13, this is possible because there cannot be a resale of the 'service' of construction.

Part Four
Broader Issues

Introduction

The last two chapters of the book have been added to this third edition. Their purpose is, firstly, to demonstrate that the theory has wider applications than discussed in the earlier part of the book. Secondly, the objective is to show that some of the theory has great practical relevance, even to some world problems. In particular, the question facing developing countries of whether to use the special characteristics of the construction industry to boost their economy, and the difficulties and consequences of doing so, is being debated by these countries and by international agencies at the present.

Chapter 16 deals with the choice of inputs in the construction process. The first part of the chapter deals with the choices facing the firm and the use of iso-product curves to illustrate the points made. The same theoretical analysis is utilised to consider the choices of developing countries referred to above. There follows a consideration of capacity of the firm, and of the industry, which is closely connected to the availability of inputs.

Chapter 17 considers the question of diversification of construction firms and the way decisions can be helped by the use of an extension of the theory described in earlier parts of the book.

16
Choice of Inputs and Capacity: The Project, the Firm and the Industry

In this chapter the choice of inputs will be considered from the point of view of the contractor and the designer and also, using some of the same pieces of theory, from the perspective of a government of a developing country wishing to utilise the construction industry in the best interests of the economy.

Inputs for the project and the firm

With each new project there is a choice as to the materials which will be used in its construction and the way it will be constructed, notably whether more labour will be used or whether it is preferred to utilise more plant and equipment. The choice is constrained at various stages of the project by the type of building the client wants, the way the building is designed, the technical possibilities of the method of construction and the costs implied in these decisions. The customary way of working does not always give adequate consideration to all these options. Indeed, the construction of a sophisticated building is so complex that it would be totally unrealistic to go back to first principles for every decision to be taken. Moreover, the traditional separation of design and construction makes it very difficult to consider alternative designs in relation to the various choices of their method of construction. This is one of the reasons for the increasing trend towards design and build as a method of procurement. There are choices where the materials used can directly affect the method of construction. The body of the building may, for example, be constructed using a steel frame, a precast reinforced concrete frame, poured reinforced monolithic concrete or load-bearing brickwork. Choices will be made according to aesthetics, ease of construction and cost of the inputs, including the labour and plant and machinery for construction. Even leaving aside aesthetic considerations, the analysis is complicated. Nevertheless, an understanding of the principles is important.

Iso-product curves*

In Figure 16.1, the choice of the combination of labour and plant and machinery, labelled 'capital', in construction of a project or of the whole output of a firm is illustrated. These are the inputs which will be considered later from the viewpoint of a government. In Figure 16.1, *AB* shows all technically possible combinations of capital and manpower which can produce 10 units of output. It is an iso-product curve. The curve has many of the characteristics of indifference curves and may be used in much the same way. Thus, point *C* shows that 10 units of output can be produced with 6 units of manpower and $4\frac{1}{4}$ of capital; point *D* that it can be produced by 5 of manpower and 5 of capital. It can also be produced by 4 units of manpower and 6 of capital at *E*, and at point *F* it can be produced by 3 of manpower and 8 of capital. It is clear that in substitution of capital for units of manpower, increasing amounts of capital are required with each unit reduction in manpower. This is known as an increasing rate of marginal substitution. There is a technical limit to the extent to which it is possible to replace manpower by capital. In order to produce 10 units of output the absolute minimum labour requirement is $1\frac{1}{2}$ units and the iso-product curve 10 becomes asymptotic to the capital axis at $1\frac{1}{2}$. There is, similarly, a minimum requirement of 1 unit of capital. There will be iso-product curves for each level of output, such as, *JK* and *LM*. (It will be noted that, unlike indifference curves, these iso-products can be given a quantitative value. It is possible to measure units of output but not units of satisfaction.) The optimum combination of inputs to produce the output of 10 units with a price relationship of 8 units of manpower costing the same as 12 units of capital and a total expenditure equal to 12 units of capital, is given at point *E*, because the price line *NP* is tangential to *AB* at *E*. Thus, at *E* the

---

* Iso-product curves can also be used in the analysis of the short-term situation by considering the effects when the amount of one input is held constant.

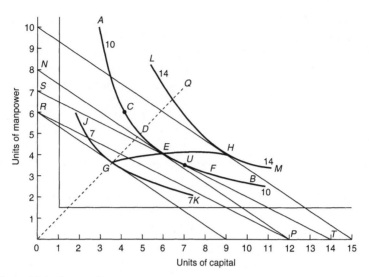

Figure 16.1 Iso-product curves

technical marginal rate of substitution is equal to the price ratio and, given the fixed total expenditure, is the combination giving the greatest possible output.

This type of analysis can be used to explain the use by firms of more capital to replace labour as the size of the business increases. Keeping the same price ratio, but changing the scale of output from *G* through *E* to *H*, the curve *GEH* is the expansion path; that is, given the prices of inputs, the way in which the firm should expand is by processes which use the combinations of inputs along *GEH*. Because the curve *GEH* has a slope lower than the slope of *OGQ*, as the firm gets bigger it pays to use a higher proportion of capital to manpower. The line *GEH* also shows that with a constant increase in expenditure from the equivalent of 9 to 12 to 15 units of capital, the increase in the product is from 7 to 10 to 14. This means that there are increasing returns to scale.

The analysis is helpful in seeing the effect of changes in the prices of inputs. Thus, if the price line changes from *NP* to *RP*, so that instead of buying 8 units of labour for the same price as 12 units of capital one can buy only 6, if the new price ratio is applied to the iso-product curve for 10 units of output by the price line *ST* parallel to *RP*, the new equilibrium point becomes *U*, with a higher ratio of capital to manpower than with the old price.

This represents what has been happening in countries with a developed construction industry. As labour has become more expensive and scarce, plant and equipment has replaced some of it. In part this may also be the result of higher transaction costs of employing labour due, for example, to more onerous employment law. An excellent example of this process of substitution was the movement in the years following the Second World War to the use of industrialised housing construction, where a major part of the dwellings was constructed in factories by mechanised processes and transported to the site for erection.

Inputs of construction for the economy

Construction in most countries is more labour-intensive than, say, manufacturing industry, so construction will generate employment directly. But that is not all. A major reason for using labour-intensive construction lies in an economic concept known as the multiplier. The expenditure multiplier is the number by which a change in expenditure must be multiplied in order to determine the resulting change in total output or gross domestic product (GDP) (Samuelson and Nordhaus, 1995). When a person is employed, he spends much of his wages on goods and services from other sectors of the economy which, in turn, generate employment and spending elsewhere, thus starting an upward spiral of increasing employment. There is therefore an employment multiplier as well. The employment multiplier may be defined as the number by which an initial change in employment must be multiplied in order to determine the resulting change in total employment in the economy as a whole.

It is tempting for governments to use the construction industry to manage the level of demand in order to reduce short-term fluctuations in the economy. In the UK the industry was used to boost demand in the 1960s and 1970s. It is, however, a difficult tool to use. One of the difficulties is the problem of timing the action by government so that the effects of the action occur at the correct time to achieve the desired objective. The lags in the construction process tend to be long and very variable. If government reduces its capital expenditure programme, it will not normally make any cuts to the projects on which construction has actually started. The build-up of the work load on a contract is slow at first, then it increases very rapidly and towards the end of the project tails off again. This pattern is similar for large projects and for small projects within the allotted duration of each project. Thus, if large projects are postponed, the reduction of the work load from what it would have been had the projects been allowed to continue would be very small for some months. If smaller projects were cancelled, the effect would be quicker but still with an initial slow effect. Later in the contract period – perhaps a year or so later – the reduction in work load would be substantial. The difficulty, therefore, about making any use of construction projects to depress demand on the economy is that, unless the government can foresee problems very far in advance, the effect will be too slow to be useful at the beginning of the period and the major effect will come much later, perhaps at a time when the contrary effect is clearly required. There is a further problem which affects the internal pressures on the resources of the industry, namely that cancellation of projects can for some time relieve pressure only on the skills and materials used at the beginning of projects. If it is, say, electricians who are in short supply, then alteration in the demands on the industry will have no effect for some months, since electricians are employed mostly at the end of any project.

At the same time, because the total process of construction from the moment the client briefs his professional adviser to the completion of the project is a long one, spanning many different occupations and skills, a sudden halt to starts of projects followed by their later commencement all together lead to disruption of the whole design process, as well as of the work done on site.

It is not surprising that international agencies, notably the International Labour Office, have been trying to find ways of stimulating labour-intensive, or 'appropriate', technologies in developing countries to raise the level of employment and GDP. There are also other advantages to the economy, such as that these technologies can usually use indigenous materials and need little plant, both of which assist the balance of payments situation. Moreover, because most developing countries have unemployment and/or underemployment, labour is likely to be relatively cheap and is largely unregulated, possibly leading to cost advantages, or at least little cost disadvantage, in using labour-intensive technologies.

It is not always easy to implement policies on appropriate technologies. The organisation of labour-intensive projects is management-intensive, which presents difficulties in poor countries. International and national donors have other agendas which often do not fit with labour-intensive methods. Attempts to persuade international aid organisations to arrange their construction projects so that they can help the development of the local construction industry, let alone boost the economy, have met with very limited success. Bilateral donors have much to gain from their funding of projects because the aid they give is usually tied to the construction being undertaken by contractors of the donor country. They, in turn, will favour high usage of capital equipment and less labour because that type of project is easier to manage than a labour-intensive one, even if the latter could be shown to be cheaper. Private investors want a particular product at as cheap a price as possible in order to make their investment worthwhile. This largely leaves the burden on governments or major industry organisations to take action (Hillebrandt, 1999b).

There are disadvantages to developing countries trying to undertake as many projects as possible by labour-intensive methods. In order to compete with international contractors within the country or elsewhere, ability to produce a sophisticated product at a competitive price requires experience in such methods and involves high productivity helped by machinery. Unless there is a part of the indigenous construction industry which is developing the skills in use in developed countries, there is a danger that the industry will never be able to compete with, and ultimately replace, expatriate contractors. Furthermore, in some countries which have, so to speak, a two-tier construction industry, with one half being run by former ex-colonials on the lines of a highly sophisticated industry and the other half run by indigenous people struggling to improve from a low level of technology, there is an urgent social and political need to integrate the two parts and raise the standards and work opportunities of the latter. South Africa is the prime example of this situation, but there are less dramatic examples in other former colonies. The conclusion reached elsewhere (Hillebrandt, 1999b) is that 'The two extreme arguments for labour-based technology on the one hand, and capital-intensive technology on the other, must, in most countries of the world, be used side by side to achieve the greatest benefit to the economy and to the construction industry.'

Capacity

The capacity of the firm is effectively determined, in the short run, by the fixed factors of production. In manufacturing industry, the most important fixed factor is usually the factory with its plant and machinery. The value of these assets will often be substantial in relation to the variable costs of production. In construction, the value of fixed assets is normally small. Moreover, the non-availability of such fixed assets would not necessarily limit the capacity of the firm. It is possible to hire plant and equipment and the administrative

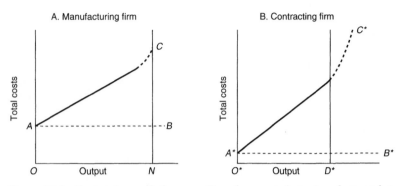

Figure 16.2 Comparison of the capacity of a manufacturing firm and a contracting firm

buildings of the contracting organisation can be overfull without limiting efficiency in the short run, especially as computer technology makes centralisation of office space less essential. It is arguable that it is management, rather than fixed assets, which is the limiting factor in the expansion of contracting firms. Whereas in manufacturing industry the short-term working of the factory is regulated by the pace and demands of the equipment, in construction the site manager has constantly to deal with new and different situations. The whole process is very decision-intensive.

Compare a manufacturing firm and a contracting firm, as illustrated in Figure 16.2. In Figure 16.2a, *AB* represents the fixed costs mainly in the form of plant and machinery. *AC* shows the variable costs of production. These rise steadily, with a slight upward tendency as the plant is used to overcapacity; that is, as repair and maintenance are postponed, leading to more breakdowns. There is, however, an absolute limit to the extent to which production can be increased when the capacity of the machinery cannot turn out any more production. Maximum output on the diagram is at *ON*.

In Figure 16.2b, the contracting firm has low fixed costs, *A*B**, because, although management is vitally important, its cost is not high, either in relation to other costs of production or in relation to the costs of the fixed assets of a manufacturer. The variable costs, *A*C**, by contrast, are high. The capacity of the contracting firm is not obvious. The existing managers can be stretched as the volume of work increases. It is very difficult to know the exact capabilities of a manager until he has been tried. If a site manager is given a project two or three times the size of any he has done before, he may rise to the challenge and perform well or he may have a nervous breakdown. If he cannot cope, that may not be immediately apparent. He may still be on site but the supervision will fail: problems will arise with subcontractors not keeping to time, making follow-on by the next team impossible in the time allocated; materials will be wasted through poor storage and pilfering; and quality will decline so that the client's representative requires reworking of some tasks. As a result, variable costs will rise very rapidly and profitability will disappear. A major difficulty is that the firm may find it very hard to judge at what output this disastrous scenario is likely to occur, because of the inherent diversity and unpredictability of human behaviour. In the figure it is at output O^*D^*, when the variable costs rise dramatically, but it could occur earlier or later.

When this scenario is combined with the fact that contractors must always bid for more projects than they want to undertake, if several of these bids are unexpectedly successful, this can lead to the disastrous situation outlined above. If substantial finance is involved, the problem is compounded.

Capacity of the industry, as necessary for any consideration of the state of the industry as a whole, is best defined rather more broadly as 'the maximum output which is attainable by the industry, within the limits of conditions considered acceptable at the time' (Hillebrandt, 1975, p. 27). There is variation as to what is acceptable, for example between war conditions and peace. Even in peacetime, what is reasonable will depend on the state of the economy and the policies of government. The supply of resources is the ultimate determinant of capacity, considered as follows:

1 the amount of a given resource in use to produce the current output of the industry;
2 the amount of the resource lying idle awaiting the demand for its use from the construction industry;
3 the extent to which the resource will increase in the time span under consideration
   (a) on the assumption of no change in intentions or policy, (b) on the assumption of effort being directed to increasing the availability of the resource and taking into account the cost of increasing its availability;
4 the extent to which the resource can be saved by substitution of some other resource and the cost of such substitution (Hillebrandt, 1975, p. 34).

This approach to industry capacity is the basis for the proposals for national resource planning for construction developed by Meikle and Hillebrandt (Hillebrandt and Meikle, 1985; Meikle and Hillebrandt, 1989).

17
Markets and Strategy

Diversification

In the previous chapters on the supply of contracting services, it has generally been assumed either that the contractor was operating in one market or that the markets were so similar that they could be considered as one. Firms' outputs are not usually confined to a single product. In the construction industry, not only do firms produce different products within the contracting operation, but they also go outside the contracting business into property development, materials production and other activities. A study of twenty large UK contractors (Hillebrandt and Cannon, 1990, p. 40) even found firms engaged in very diverse activities having virtually no relation to construction, including manufacture of boats, computer software and gold mining.

In many cases, firms have diversified into other businesses for quite specific reasons. Cannon and Hillebrandt (1989, p. 31) define diversification as the process by which firms extend the range of their business operations outside those in which they are currently engaged. This broad definition includes (a) the process referred to as backward vertical integration, that is, the acquisition or development of businesses whose products are inputs to the firm's own main operations; (b) forward integration, that is, the extension of the firm's activities to those of the normal purchase of its products; (c) horizontal diversification, that is, a movement into other markets not involving the firm in any vertical relationships as in (a) and (b) above. Any of these forms of expansion may take place either by internal development, or by merger, or takeover.

One of the main reasons for backward vertical integration in the construction industry is uncertainty in the availability of supplies, especially during the periodic booms in the industry. A firm which controls its own source of supply is more likely to be able to meet delivery schedules. It also avoids the transaction costs of purchasing, including price negotiation. Such vertical integration is most likely to be beneficial if the material concerned is a major input to a project and one where transport costs are high. For these reasons, road builders often own and manage their source of aggregates from a site as near as possible to the location of the project.

Horizontal integration may simply consist of merging with another firm in the same business. This is a means of growth. A firm may link up with a firm in a business which is in some way connected to the main business. There may be common resources, in which case there may be economies of scale in purchasing or it may be possible to share certain resources. Lastly, a firm may link with a firm which has no connection with the original business. The reason for this action is normally a desire to spread risk. In the construction industry, contracting organisations have been interested in buying businesses which were not greatly affected by trade cycles, so that, if the contracting business was in a cyclical low, the other business would carry the whole organisation until construction recovered.

In the UK in the 1980s the fashion for diversification left some contracting firms with a mixed bag of firms. In some cases, while there had been good reason to make a particular acquisition, the reason was no longer relevant. When the recession came, there was need to reappraise the portfolio of businesses. Unfortunately, at that time there were urgent decisions to be made and firms often divested themselves of businesses that were easy to sell, rather than following a logical long-term strategy. There are a number of principles on which logical decisions can be taken on which markets to be in, and the most appropriate balance of resources to be invested in each.

Allocation of a firm's resources

Assuming, for the moment, that the construction firm wishes to maximise profit, a simple approach is to so arrange the deployment of the firm's resources and hence output that a redistribution between businesses cannot increase the total profit. Consider three businesses. Profit in each business is maximised when marginal revenue equals marginal cost. This is the same thing as saying that marginal profit from any additional unit of output is nil and will become negative if output is increased (see Figure 15.1). Suppose,

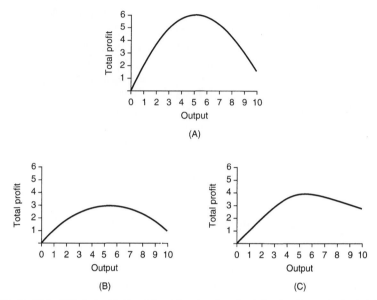

Figure 17.1   Total profit at various levels of output for business A, B and C

however, that a change in circumstances, such as a credit squeeze or the loss of one or more managers, means that the firm has not the level of resources to sustain the three businesses at the optimum levels of output: what should the firm do? The answer must, to some extent, depend on how permanent the change in circumstances is. It may pay to reduce the scale of activity in each business if that is for a short time only. If the change is long-term, then the total profit in each business must be considered and the most profitable kept. It may not be a clear-cut decision, however.

In Figure 17.1 units of output are measured on the horizontal axis and on the vertical axis the total amount of profit at each output. The output of maximum profit is 5 units each for businesses A, B and C, that is 15 units in all, with a profit of 13 units. If output has to be reduced to 10 units, the firm should keep businesses A and C but get rid of B, leaving a profit of 10 units. If the output has to be reduced to 7 units of output, business A should be kept, but it is unclear what should happen otherwise. The profit of 2 units is the same from running business B or business C at 2 units of output or both B and C at one unit each, giving a total profit of 8 units from business A plus either B or C at 2 units of

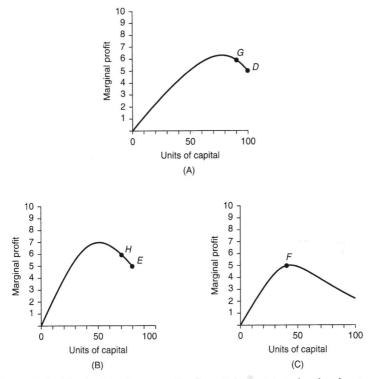

Figure 17.2 Marginal profit per unit of capital at various levels of output for businesses A, B and C

output or B and C at one unit of output each. For that matter, one could lose only 0.2 units of profit by reducing the output of A to 4 units and increasing the output of B or C. Clearly, other factors have to be considered.

One of the ways of determining the appropriate business to concentrate on is to consider which are the critically scarce resources. For many businesses, capital is the principal limiting factor in expansion or, as in the example above, in curtailing activities. This is true of construction to a large degree in many countries, especially where there is no developed capital market and where working capital is not provided, as it is in the UK, by other participants in the process. In construction, however, it is arguable that management is the principal limiting factor and, as is argued in Chapter 16, this would seem to apply both in developed and in developing countries. Whatever the scarce, most limiting factor, it is instructive to look at the marginal returns to this scarce factor of production as a way of considering the businesses in which best use may be made of it. Consider the three firms A, B and C again, in Figure 17.2, and their use of the scarce resource.

On the horizontal axis are the units of the scarce resource and on the vertical axis the marginal profit made by an additional unit of the resource. We assume for ease of exposition that the scarce resource is capital. If the transfer of one unit of capital from one business to another increases the marginal profit, then that is worthwhile. The equilibrium position is where the marginal returns to a unit of capital are as high as possible and equal in all businesses, because in that situation no improvement can be made by a transfer. More specifically, a marginal profit of 5 units can be made on units of capital, with deployment of 100 in business A at D, 80 in business B at E and 40 in business C at F, making 220 in all. Suppose it is necessary to cut back the units of capital by 60 units, to 160 units. If the amount of capital in business C is reduced, the marginal return will fall, whereas the return in businesses A and B will rise. As the marginal rate of return must be the same in all businesses, the optimum solution is to employ 90 units of capital in business A at G and 70 in business B at H, giving a marginal profit in both of 6 units. There is no problem in applying this logic to the use of finance which is essentially homogeneous. However, its application to the scarce resource of management is not so practical because of the specific management skills which may be required in some businesses. However, the principle is clear.

Other factors have traditionally been considered by management specialists, though they are very relevant to the economic study of the firm. They include annual market growth rate, relative competitive position, business strengths and market prospects (Ramsay, 1989). In discussion of these measures Hillebrandt *et al.* (1995) point out that, in the UK recession of the early 1990s, the businesses which were disposed of were not usually those with low ratings of profitability or other measures, because those businesses would realise little at a time when companies were short of cash. They also point out the implicit assumption in much of the management literature that profits are positive and hence that large turnover is a 'good thing'. This is clearly not so if the businesses with large turnovers are making losses on that turnover. This is one of the areas where stronger links between management and economic theory would be beneficial.

Appendix A: Indifference Curve Analysis Applied to the Demand for Housing

In Chapter 4 the demand for housing is considered, commencing with the concept of the individual's demand curve for units of housing. In this appendix the individual's demand curve is derived from his indifference curves.

The demand for housing depends initially on the consumer's assessment of the desirability of housing compared with all the other goods and services he could buy. Before embarking on indifference curves for housing, let us consider the general case of two commodities.

In Figure A.1, curve I_1 is an indifference curve showing all the various combinations of commodity A and commodity B between which the consumer is indifferent or, to put it another way, which give him equal satisfaction. The consumer does not mind, for example, whether he has 15 units of commodity B and 3 units of commodity A, as at point C, or 4 units of commodity B and 10 of commodity A as at point D. If both commodities are desirable goods, indifference curves are convex to the origin as shown in Figure A.1. Generally speaking, the more a person has of a commodity, the less value is placed on additional units of it. Thus to give up one unit of commodity B when he has only 4 units, the consumer needs 4 units of commodity A to compensate him. But when he has 15 units of commodity B, he needs only half a unit of commodity A to compensate him for the loss of one unit, i.e. to remain on the same indifference curve.

There will be an indifference curve for each level of consumer satisfaction. The higher the indifference curve, or the further away from the origin, the greater the level of satisfaction. In Figure A.1, in moving from indifference curve I_1 to indifference curves I_2 and I_3, the consumer is able to have the same amount of one commodity and at the same time increase the amount of the other. Thus at E he has 20 units of commodity B and $2\frac{1}{2}$ of commodity A; at F on indifference curve I_2 he still has 20 units of commodity A but in addition has 6 of A, that is $4\frac{1}{2}$ more than on indifference curve I_1; and at point G he has the same amount of B with 15 of A. He is clearly better off at G than at F and at F than at E. In fact each consumer has an infinite number of indifference curves forming an indifference map, each curve being a contour on the hill of satisfaction. We cannot assign values to the contours because it is not possible to say that the consumer is, say, twice better off on a certain indifference curve than on another. All we know is that he becomes progressively better off as he goes 'up the hill'.

Going from the general case to the particular, we can put housing on one axis and all other commodities on the other axis and consider the consumer's indifference map for housing and all other commodities. For

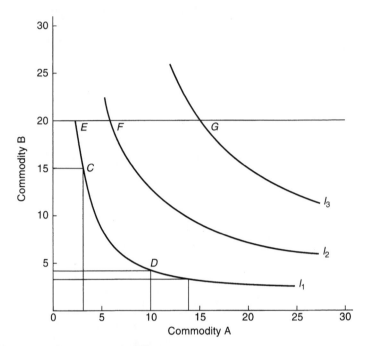

Figure A.1   A consumer indifference map

present purposes, the simple assumption (discussed in Chapter 4) is made that one can somehow on a points system conceive of homogeneous units of housing, and these housing units are measured on the horizontal axis in Figure A.2 as notional housing units consumed per annum. It is in practice difficult to measure 'all other goods' except in terms of money, and therefore on the vertical axis all other goods are in money terms as annual income.

What can be said about the housing indifference map? First, the general shape of the curve will follow the usual indifference curve with two desirable goods. Housing, however, is special in that some minimum units of housing are very desirable and it is therefore likely that the curve with a low level of satisfaction will become asymptotic to a line parallel to the vertical axis. In other words, no matter how much income is given in compensation, there is a minimum standard of housing required. In Figure A.2 the indifference curves are shown as being asymptotic to a line parallel to the vertical axis, showing that the minimum number of units of housing at the level of satisfaction shown by indifference curve I_1 which are acceptable is 15. Similarly, the minimum income desired or other essential purposes such as food is shown as equivalent to 50 units of income per annum. The area between the axes and the lines *CD* and *DE* may in some sense be regarded as the area in which the standard of living is below acceptable

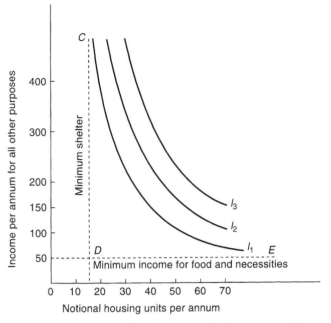

Figure A.2 Indifference curves for housing and other commodities

social standards, at least as seen by the individual to whom these indifference curves refer.

Indifference curves for housing and the price of housing

The reaction of the individual consumer, with his scales of preference illustrated in his indifference map, to the market situation must now be examined.
Consider first the price of housing. This may be indicated by a price line
showing the rate at which income can be exchanged for housing, as shown
by, for example, *AB* in Figure A.3. Along *AB* a constant expenditure will
'buy' the proportions of housing and income per annum as shown by the
abscissae of any point. Thus he may have *OA* of income and no housing,
or *OB* of housing and no income, or any combination in between – such
as *DC* of housing and *CE* of income. *AB* represents his opportunities. He
will be as well off as he can be if, given this price relationship and income
OA, he gets on to the highest indifference curve possible, and this will be
that indifference curve which is tangential to the price line *AB*. His optimum position will be at *C* where he will choose to have *OD* of income and
OE of housing. The consumer has insufficient income to be at any other
point than *C* on this indifference curve. If he were on another indifference

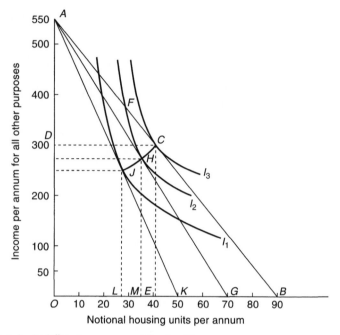

Figure A.3   Indifference curves and price lines

curve, say at *F*, which is within his income, he would be better off by moving along the price line *AB* to *C*, because he is then on a higher indifference curve and therefore better off.

If the price of housing changes, the level of income remaining the same, the consumer will adjust so that at the new level of prices he is at his optimum position. If the price rises to *AG* the consumer will move from *C* to *H*. If it rises again to *AK* he will move to *J*.

The optimum points with each fresh price may be joined to form a price consumption curve *JHC* (sometimes known as an offer curve), showing how much will be consumed at each price.

Changes of income

At least as interesting as the change in the amount consumed at each price is the effect of a change in the consumer's income. This is particularly so as housing expenditure usually takes a large proportion of total income. In Figure A.4 the increase is shown as a movement from *AB* to *CD* to *EF* in which the price relationship, shown by the slope of the lines, is constant and hence the income lines are parallel.

As income increases from *OA* to *OE*, the consumer changes from demanding *OG* housing and keeping *OJ* income to *OH* housing and keeping *OK* in-

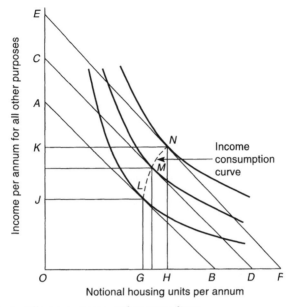

Figure A.4 Indifference curves and income lines

come. An income consumption curve (sometimes known as an Engel curve) joining the points of optimum combination with varying levels of income may be obtained – *LMN* in Figure A.4.

The income consumption curve normally slopes upwards to the right. There are, however, certain goods, known as inferior goods, of which the consumer demands less as his income rises because he replaces them by more luxurious commodities, and in this case the income consumption curve slopes upwards to the left. This situation could theoretically apply to housing if, for example, very rich people chose to live in hotels or yachts rather than in a normal dwelling. It is certainly unlikely to apply for large groups of individuals.

Effect of price change: part income and part substitution effect

The effect of a price change may be further analysed into (*a*) the result of the price change of the commodity on the real income of the consumer, and (*b*) the result of the price change in altering the relative prices of commodities.

If the price of housing falls, the consumer has in fact had an increase in his real income. This is illustrated in Figure A.5. *AC* is the original price line with income of *OA*. This is tangential to the indifference curve I_1 at *F*

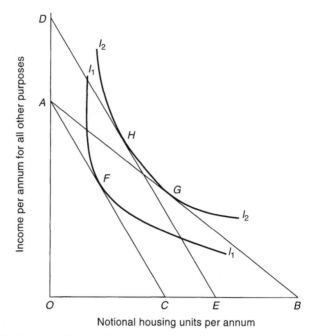

Figure A.5   Income effect and substitution effect

and *F* is the optimum point. Price then falls to *AB*. The new optimum position is *G* on indifference curve I_2. If the price had not changed, the consumer could have arrived at the indifference curve I_2 at *H* by an increase in income of *AD*, for *DE* is the price line parallel to *AC* representing this increase in income. The fall in price has given the consumer an increase in his real income of *AD*. In fact, because price has changed the consumer wishes to substitute the cheaper commodity for others and hence moves from *H* (the equilibrium point if income had increased with no price change) along his indifference curve I_2 to *G*.

The demand curve of the individual consumer

It is possible from the price consumption curve illustrated in Figure A.3 to derive the consumer's demand curve showing how much of a commodity he will demand at various prices. Thus, deriving Figure A.6 from Figure A.3 at a price given by the slope of *AK*, that is, *AO/OK* = 550/50 i.e. 550 units of income for 50 units of housing or 11 units of income for 1 unit of housing, the amount demanded is *OL* = 27 units of housing; similarly, at a price given by the slope *AG*, i.e. 550/70 or 7·9 units of income for 1 of housing, the amount demanded is *OM* = 35 units; at a price given by slope

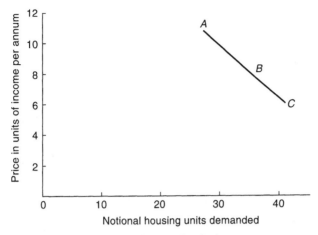

Figure A.6 The demand curve of the individual

AB, that is, 550/90 or 6·1 units for 1 unit of housing, the amount demanded is *OE* = 41 units.

There is a further point to note, namely the relation of the price consumption curve to the demand curve of varying elasticities.

The price consumption curve shows the amount of income which will be devoted to a given commodity at various prices. In Figure A.3 the price consumption curve rises as the price falls. This means that the consumer gives up less and less of his income with each fall in price – in other words, demand is inelastic. If the price consumption curve is falling, then demand is elastic, and if it is horizontal, the demand has an elasticity of 1.

The effect of capital on the demand for housing

In the theoretical indifference curve analysis, homogeneous housing units have been considered as an alternative to all other goods (expressed as income per annum for all other purposes) and the price relationship is then expressed in terms of the price of housing per annum. The effect on demand of changes in income has been considered. In practice, housing may be rented or it may be bought, either with owned or borrowed capital. Most other goods – as represented on the vertical axis – are bought out of income rather than out of capital. Housing is a durable good with a high resale value and is therefore considered as a capital good.

There seem to be three main ways of dealing with the effect of capital ownership on the demand for housing:

(*a*) To do nothing: as income may be earned or derived from capital investments, to consider income as an alternative to capital is already including some allowance for the capital component.

(*b*) To convert capital at the 'annuity rate': in order to increase the allowance for capital to its proper value in the decision to purchase a house over, say, twenty years, the capital could be converted to income at the 'annuity rate', that is, capital depreciating to nil at the end of the period. This would conform to building societies' attitudes, although the allowed period of repayment is now greater than it was. This method requires knowledge of capital in private hands. There are now regular statistics for the UK of the estimated wealth of individuals, based on the various wealth taxes collected by the Inland Revenue and on other sources, broken down by types of wealth, including UK residential buildings, other UK buildings, various types of securities, superannuation benefits, insurance policies, and so on (Inland Revenue, 1998, Table 13.3).

(*c*) To change the indifference map: the effect of capital is probably in any case not measured only by a discounting technique because a person with capital has more security and will therefore have a different attitude to committing a large amount of money. The best solution may therefore be to consider income as in (*a*) above and in addition acknowledge capital ownership as a factor which will change the shape of the indifference map and shift the demand curve to the right.

Appendix B: Degree of Potential Surprise Analysis of Tendering

In Chapter 14 four stages of the contractor's response to an invitation to bid are identified: the decision to bid or not to bid, determination of the cost of the project, assessment of the lowest worthwhile bid price and the determination of the final bid price. In this Appendix the last three of these are analysed, using the degree of potential surprise as developed by Shackle in a number of publications (for example, 1952 and 1955). Decisions on the cost of undertaking the contract and the determination of the final bid price have a high degree of uncertainty and the decision on the lowest worthwhile bid price is closely related to the state of the business and the character and personality of the contractor. Shackle's approach is particularly suited to these types of problem. It has been adapted here to the specific contracting industry tendering situation.

Cost of undertaking the project

The degree of potential surprise is the surprise or shock which would be felt by a person at the occurrence of an event about which uncertainty had existed. This is applied to the assessment of the uncertainty of the cost estimates of a project. In Figure B.1, on the horizontal axis are measured the losses or profits which may be made with a given bid. On the vertical axis is measured the degree of potential surprise which would be registered by the entrepreneur at each of these possible losses and profits. The degree of potential surprise ranges from no surprise to the greatest surprise which can be experienced; that is, from certainty to impossibility. An arbitrary scale of 0–10 has been adopted. Thus, the entrepreneur regards it as impossible (degree of potential surprise of 10) that bid $x + 5$ (x being his estimated cost and 5 his percentage profit) will produce a loss as high as 10 per cent or a profit of 15 per cent, but perfectly possible, i.e. would not be surprised at all (hence degree of potential surprise of 0), if it produced anything from a loss of 2 per cent to a profit of 7 per cent. At ranges between 2 per cent and 10 per cent loss and between 7 per cent and 15 per cent profit, he would experience varying degrees of potential surprise.

The concept of degree of potential surprise is much less precise than that of probability analysis, but probability analysis is not applicable to this particular situation because most of the variable factors are uncertainties to which no probability can be assigned and of which the assessment is largely qualitative, not quantitative.

The advantages of the degree of potential surprise are that (a) it is applicable to situations not susceptible to probability analysis; (b) it recognises that for the entrepreneur a number of outcomes are all perfectly and equally

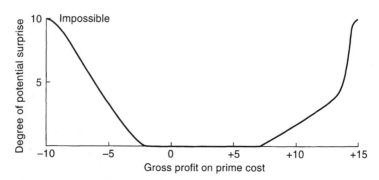

Figure B.1   Degree of potential surprise at a bid *x* + 5 making levels of profit or loss

possible; (*c*) if the possibility of one event happening comes to seem more likely than before, it does not imply that all others become less likely; (*d*) it does not give any impression of being pseudo-accurate as subjective probability tends to do; and (*e*) it is probably more akin to the way the entrepreneurs in the construction industry actually think, with the possible exception of a few sophisticated contractors.

The stimulus of various outcomes

In practice, although he may acknowledge all these possibilities, no decision-maker can keep in his mind the whole range of outcomes at the same time, and he will probably tend to concentrate his attention on one possible level of profit and one possible level of loss with their relevant degrees of potential surprise. These will stimulate his imagination. He will fix on them just as a man filling in the football pools will choose, according to his temperament, to go for small gains with high chances of obtaining them or large gains with tiny possibilities of realisation or somewhere in between. These will be the examples of potential gain or loss which would be used in board discussions. In order to understand the theoretical basis of the selection of the points of stimulus, Shackle's concept of stimulus indifference curves can be used with special application to the bid situation.

A stimulus indifference curve is the locus of points where the stimulus to the imagination of the entrepreneur afforded by a combination of surprise and profit or loss is constant. In Figure B.2 these stimulus indifference curves are shown by the dotted lines. The decision-maker will receive equal stimulus on, say, indifference curve *AB* from a profit of 5 per cent with no degree of potential surprise and a profit of 10 per cent with about 5 points of degree of potential surprise and at all points along the curve. He will be more stimulated – and therefore be on a higher indifference curve – by a higher profit of, say, 8 per cent with no degree of potential surprise (point *C* on indifference curve *CD*). Similarly, he will have stimulus indifference curves relating degree of potential surprise to loss.

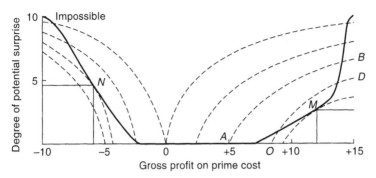

Figure B.2 Stimulus indifference curves and the degree of potential surprise at outcome of bid $x + 5$

This stimulus indifference map may be superimposed on the degree of potential surprise curves (Figure B.1), as it has been in Figure B.2. The point on which the entrepreneur will focus his attention will be that giving him the greatest stimulus, i.e. where his highest stimulus indifference curve is tangential to the degree of potential surprise curve, at N and M in Figure B.2. In this case the highest stimulus would be obtained by a profit of 12 per cent combined with a degree of potential surprise of $2\frac{1}{2}$ and a loss of 6 per cent with a degree of potential surprise of $4\frac{1}{2}$.

In order to standardise the degree of potential surprise in comparing profit and loss, the entrepreneur can move along the stimulus indifference curve to nil degree of potential surprise. The value of the profit with this nil degree of potential surprise is called the standardised focus gain, and the corresponding value of the loss with nil degree of potential surprise is called the standardised focus loss. The standardised focus gain would then be 9 per cent profit and the standardised focus loss 4 per cent.

Lowest worthwhile bid price

The gambler indifference map

How can a businessman decide, when he has in mind a perfectly possible profit of 9 per cent and a perfectly possible loss of 4 per cent, whether or not he would go in for the contract at that price? Clearly, it will depend on his personal characteristics – how far he is prepared to gamble a gain for a loss. It will also depend on the state of the business. Since one of the features of construction is that any one contract forms a relatively large part of the firm's annual turnover, the decision on any one contract is an important determinant of the profit or loss for the year. Hence the decision-maker cannot rely on gains and losses offsetting each other in the period relevant to him (if he could, he could simply say a standardised focus gain of x per cent is greater than the standardised focus loss and therefore it would be worthwhile at that price).

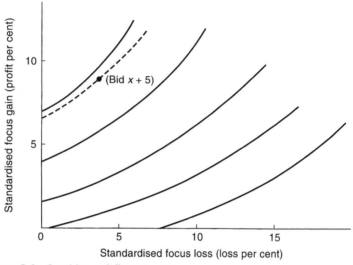

Figure B.3 Gambler indifference map

Moreover, his reaction to the chances of loss or gain will be to some extent determined by other objectives of the business. The entrepreneur may, for instance, wish to obtain a particular contract because it carries personal prestige. If he has a contract manager whom he especially wants to keep and for whom this job is eminently suitable, he may be more prepared to stand a loss.

It is unlikely in practice that the process of estimating costs overtly takes into account the position of the firm on its cost curve. However, the entrepreneur does consider his cost function explicitly or implicitly after the initial assessment of costs in the decision on what mark-up to apply if the firm is on the downward slope of its cost curve. If the overheads are not spread over a sufficient volume of work, the entrepreneur will be more prepared to take the work than if the firm's capacity is becoming fully stretched.

Two individuals may react quite differently to the same situation. If, to take an extreme example, the company is on the verge of insolvency, one man may be prepared to risk all on one large contract which could pull the firm through completely but could be disastrous, while another man will hope for a number of less risky jobs and be prepared to wait years for financial viability rather than become insolvent.

In short, each decision-maker will have his own gambler indifference map, as suggested by Shackle, reflecting partly his own personality and partly the state of the business. It may change with every change in the state of the business and hence with every new contract won. Such a gambler indifference map is shown in Figure B.3. The indifference curves take the shape to be expected with two variables, one of which is desirable and the other undesirable.

The values of the standardised focus gain and the standardised focus loss may be plotted on the gambler indifference map. If the point falls on an indifference curve cutting the gain axis, the project is worthwhile to the firm at that price. If it falls on an indifference curve cutting the loss axis, it is not worthwhile. The standardised focus gain for bid $x + 5$ of 9 per cent and the standardised focus loss of 4 per cent would fall on an indifference curve cutting the gain axis at $6\frac{1}{2}$ per cent. This value might be called the ultimate focus gain, since it is that derived after taking into account both the stimulus indifference curve and the gambler indifference curve.

All this analysis of bid price in relation to the firm has been undertaken in respect of one bid price, $x + 5$. Similar analysis can be undertaken for prices above and below $x + 5$, i.e. $x - n, \ldots, x - 3, x - 2, x - 1, x + 1,$ $x + 2, x + 3, \ldots, x + n$. For each bid price there will be an ultimate focus gain or loss and each can be plotted on the one gambler indifference map.

Towards winning a profitable contract

Likelihood of getting the job

Within the firm, the higher the mark-up, the higher the ultimate focus gain. It is also clear that the higher the bid price, the lower the likelihood of getting the job at all. For each bid price which the firm wishes to consider (those which lead to an ultimate focus loss can be eliminated) there will be an ultimate focus gain and an assessment of the likelihood of obtaining the job, which may be expressed in terms of degree of potential surprise. This should take into account as much information as possible on the history of the firm in competing against various competitors. It should include past tender data, an assessment of the overall state of the market, any information possible on the level of competitors' costs compared with own costs, including any advantage in purchasing or subcontracting, an idea of their current work load and hence their position on their own cost curves, and how far they are likely for other reasons to want the job.

Bidding indifference map

Just as the decision maker had a gambler indifference map balancing loss against gain, so he will have an indifference map balancing degree of potential surprise of getting the contract against ultimate focus gain if he does. This might be named a bidding indifference map.* Its shape will be governed principally by the state of the firm's order books, an assessment of how the market price is likely to change in the future and of contracts coming on to the market in the future. Figure B.4 illustrates the bidding indifference map of a contractor motivated to maximise profits.

For each bid price the ultimate focus gain and degree of potential surprise can be plotted as AB in Figure B.4. Bid $x + 5$, for example, is plotted at D with ultimate focus gain of $6\frac{1}{2}$ per cent and degree of potential surprise of $1\frac{1}{2}$ (on a scale 0–10). The curve of bid prices on this map may be very varied in shape, since it is determined by so many factors, but it may

---

* The analysis here is a further development from that of Shackle.

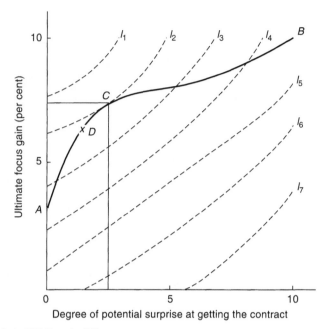

Figure B.4   Bidding indifference map

retain some of the shape of the normal degree of potential surprise curve. Line *AB* is tangential to the highest indifference curve at *C*, giving an ultimate focus gain of $7\frac{1}{2}$ per cent and a degree of potential surprise of obtaining the contract of $2\frac{1}{2}$. This will correspond to a bid price substantially higher than $x + 5$.

Alternative objectives of the firm

If the entrepreneur wished to obtain only normal profit and thereafter to maximise turnover, his strategy would be somewhat different. He would need to obtain on each contract a minimum level of profit, and the contractor would tender or negotiate in order to get as much work as possible at this relatively low profit, irrespective of what other job opportunities were coming up which he might be able to obtain more profitably. His bidding indifference map will show that he values a low degree of potential surprise of getting the contract highly in relation to the ultimate focus gain, as he is not so interested in gains as long as they are positive. The situation would be as illustrated in Figure B.5 below. If the indifference curves with a profit maximiser are I_1, I_2, I_3, etc. (as in Figure B.4), with a revenue maximiser (with a minimum profit constraint) they will change to, say, J_1, J_2, J_3. This must be so because, considering the point *D* where

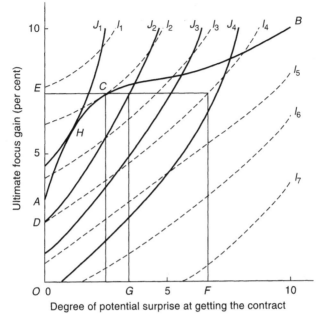

Figure B.5 Bidding indifference map: alternative assumptions

the two series intersect on the ultimate focus gain axis, a person who wishes to maximise profit would, for increase in ultimate focus gain of *DE*, be prepared to have a degree of potential surprise of *OF*, while a person who is not very interested in profit would certainly wish to achieve a much smaller degree of potential surprise – *OG* – to obtain an ultimate focus gain increase of *DE*. Thus the optimum point, assuming the same bid-price curve as *AB* in Figure B.4, will be at *H*, with a lower ultimate focus gain combined with a lower degree of potential surprise for the revenue maximiser than for the profit maximiser.

Value of the analysis

Thus this analysis tries, firstly, to develop a theory of the stages of thought-processes through which entrepreneurs go, consciously, or unconsciously, in coming to a decision on tendering, and secondly, by bringing these stages and a possible method of analysis to the forefront of the entrepreneur's mind, it tries to provide a framework to help the more sophisticated entrepreneur to analyse and improve his decisions. It is not suggested that entrepreneurs should now spend their time drawing their degree of potential surprise curves and their stimulus, gambler and bidding indifference maps, although it would probably be possible for entrepreneurs to indicate where they were on such maps. It is, however, suggested that the analysis would be helpful in understanding the reason for decisions on tender prices; in locating the reasons for differences of opinion between persons sharing the entrepreneurial function within contracting firms; and in making the process of tender decisions more logical and efficient.

References

Ahmad, I. and Minkarah, I. (1988), 'Questionnaire Survey on Bidding in Construction', ASCE, *Journal of Management in Engineering Divisions*, vol. 4, no. 3, July, 229–43.

Aikivuori, A (1996), 'Periods of Demand for Private Sector Housing Refurbishment', *Construction Management and Economics*, vol. 14 no. 1, 3–12.

Akintoye, A. and Skitmore, M. (1991), 'Profitability of UK Construction Contractors', *Construction Management and Economics*, vol. 9, no. 4, July, 311–25.

Archibald, G.C. (1987), 'Firm, Theory of the', in J. Eatwell, M. Milgate and P. Newman (eds), *The New Palgrave: A Dictionary of Economics*, Vol. 2 (London: Macmillan).

Audit Commission (1997) for Local Authorities and the National Health Service in England and Wales, *Rome Wasn't Built in a Day,* Audit Commission.

Ball, M. (1988), *Rebuilding Construction: Economic Change in the British Construction Industry* (London: Routledge).

Ball, M. and Grilli, M. (forthcoming), 'Contestability and the Persistence of Profits in the UK Construction Industry', *Construction Management and Economics*.

Baumol, W.J. (1959), *Business Behavior, Value and Growth* (New York: Macmillan).

Baumol, W.J., Panzar, J.C. and Willig, R.D. (1982), *Contestable Markets and the Theory of Industry Structure* (New York: Harcourt Brace Jovanovich).

BCIS (Building Cost Information Service) (1999), *Indices and Forecasts* (London: BCIS).

Benjamin, N.B.H. (1969), 'Competitive Bidding for Building Construction Contracts', Technical Report No. 106, Construction Institute, Department of Civil Engineering, Stanford University.

Betts, M. and Lansley, P. (1993), '*Construction Management and Economics*: A Review of the First Ten Years', *Construction Management and Economics*, vol. 11, no. 4, July.

Bowley, M.E.A. (1960), *Innovations in Building Materials* (London: Duckworth).

Bowley, M.E.A. (1966), *The British Building Industry* (Cambridge: Cambridge University Press).

Broemser, G.M. (1968), 'Competitive Bidding in the Construction Industry', PhD dissertation, Stanford University.

Buckley, P.J. and Enderwick, P. (1989), 'Manpower Management', in P.M. Hillebrandt and J. Cannon (eds), *The Management of Construction Firms: Aspects of Theory* (London: Macmillan).

Building (1998), 'The Building Top 50', *Building*, 17.7.98, 38–43.

Building (1999), 'The Building Top 50', *Building*, 23.7.99, 43–8.

Building Maintenance Committee (1972), *The Report of the Committee on Building Maintenance* (London: HMSO).

Cannon, J. and Hillebrandt, P.M. (1989), 'Diversification', in P.M. Hillebrandt and J. Cannon (eds), *The Management of Construction Firms: Aspects of Theory* (London: Macmillan).

Cavill, N. (1999), 'Big in Japan', *Building*, issue 22, 4 June, 24–5.

CFR (Construction Forecasting Research, Ltd) (1999), *Construction Forecasts 1998–1999–2000*, A Report by the Joint Forecasting Committee for the Construction Industries (London: CFR).

Cheshire, P. and Sheppard, S. (1998), 'Estimating Demand for Housing, Land and Neighbourhood Characteristics', *Oxford Bulletin of Economics and Statistics*, vol. 60, no. 3, 357–82.

CIB TG29 (1998), 'First meeting of the Task Group, Construction in Developing Countries', Arusha, 21–3 September.

CIC (Construction Industry Council) (1998), *Constructors' Guide to PFI* (London: Telford for CIC).

CIOB (Chartered Institute of Building) (1982), *Project Management in Building* (Ascot: CIOB).

Clark, C. and Jones, G.T. (1971), *The Demand for Housing*, University Working Paper No. 11 (London: Centre for Environmental Studies).

Coase, R. (1937), 'The Nature of the Firm', *Economica*, vol. 4, 386–405.

Colean, M.L. and Newcomb, R. (1952), *Stabilizing Construction: The Record and Potential* (New York: McGraw-Hill).

Cyert, R.M. and March, J.G. (1963), *A Behavioral Theory of the Firm* (Englewood Cliffs, New Jersey: Prentice-Hall).

DETR (1995), *Projection of Households in England to 2016* (London: HMSO).

DETR (1998a), *Rethinking Construction* (The Egan Report) (London: DETR).

DETR (1998b) *Housing and Construction Statistics: Great Britain* (London: The Stationery Office).

Dixon, I. (1990), 'The Management of Building Maintenance: A Builder's Perspective', in I.K. Quah (ed.), *Building Maintenance and Modernization*, Vol. 1, Proceedings of the International Symposium on Property Maintenance, Management and Modernization, CIB Working Commission 70 (Singapore: Longman).

DLC (Davis, Langdon Consultancy) (1997), 'Unpublished Estimates of World Construction Output'.

DL&E (1999), *Contracts in Use: A Survey of Building Contracts in Use during 1998* (London: RICS).

DL&E (2000), *European Construction Costs Handbook* (London: Spons).

DLSI (Davis Langdon & Seah International) (1995), *Construction and Development in Vietnam* (Hong Kong: DLSI and London: DLC).

DLSI (Davis Langdon & Seah International) (1997), *Asia Pacific Construction Costs Handbook* (London: Spons).

Drew, D. and Skitmore, M. (1997), 'The Effect of Contract Type and Size on Competitiveness', *Construction Management and Economics*, vol. 15, no. 5, 469–89.

Drewer, S. (1999), 'That Shambling Phenomenon Called Construction', *Habitat International*, vol. 23, no. 2, 167–76.

Ermisch, J.F., Findlay, J. and Gibb, K. (1996), 'The Price Elasticity of Housing Demand in Britain: Issues of Sample Selection', *Journal of Housing Economics*, vol. 5, 64–86.

Flanagan, R. (1990), 'Making International Comparisons in the Global Construction Market', Proceedings of the CIB W 55/65 Symposium, *'Value in Building Economics and Construction Management'*, Sydney, March.

Flanagan, R. and Norman, G. (1989), 'Pricing Policy', in P.M. Hillebrandt and J. Cannon (eds), *The Management of Construction Firms: Aspects of Theory* (London: Macmillan).

Friedman, L. (1956), 'A Competitive Bidding Strategy', *Operations Research*, vol. 4, 104–12.

Friedman, M. (1957), *A Theory of the Consumption Function* (Princeton: Princeton University Press).

Gates, M. (1967), 'Bidding Strategies and Probabilities', *Journal of the Construction Division ASCE*, vol. 93, no. C01, March, 75–103.

Gates, M. (1983), 'A Bidding Strategy Based on ESPE', *Cost Engineering*, vol. 4, no. 6, December, 27–35.

Goodman, A.C. (1989), 'Topics in Empirical Urban Housing Research', in R.F. Muth and A.C. Goodman (eds), *The Economics of Housing Markets* (Chur: Harwood Academic Publishers).

Green, S.D. (1999), 'The Missing Arguments of Lean Construction', *Construction Management and Economics*, vol. 17, no. 2, March.

Grinyer, P.H. and Whittaker, J.D. (1974), 'Management Judgement in a Competitive Bidding Model', *Operational Research Quarterly*, vol. 24, 181–91.

Groak, S. (1992), *The Idea of Building: Thought and Action in the Design and Production of Buildings* (London: Spon).

Groak, S. (1994), 'Is Construction an Industry?', *Construction Management and Economics*, vol. 12, no. 4, July, 287–93.

Groenewegen, P. (1987), '"Political Economy" and "Economics"', in J. Eatwell, M. Murray and P. Newman (eds), *The New Palgrave: A Dictionary of Economics*, Vol. 4 (London: Macmillan).

Gruneberg, S.L. and Ive, G. (2000), *The Economics of the Modern Construction Firm* (London: Macmillan).

Hague, D.C. (1969), *Managerial Economics; Analysis for Business Decisions* (London: Longman).

Hendershott, P.H. and Shilling, J.B. (1982), 'The Economics of Tenure Choice, 1955–79', in C.F. Sirmans (ed.), *Research in Real Estate*, Vol. 1 (Norwich, CN: JAI Press).

Hillebrandt, P.M. (1975), 'The Capacity of the Industry', in D.A. Turin (ed.), *Aspects of the Economics of Construction* (London: Godwin).

Hillebrandt, P.M. (1984), *Analysis of the British Construction Industry* (London: Macmillan).

Hillebrandt, P.M. (1999a), 'Problems of Larger Local Contractors: Causes and Possible Remedies', paper delivered at the Second Meeting of CIB Task Group (TG 29), Construction in Developing Countries – Contractor Development, Kampala.

Hillebrandt, P.M. (1999b), 'Choice of Technologies and Inputs for Construction in Developing Countries', paper presented at 2nd International Conference on Construction Industry Development: Construction Industry Development in the New Millennium, Singapore, October.

Hillebrandt, P.M. and Cannon, J. (eds) (1989), *The Management of Construction Firms: Aspects of Theory* (London: Macmillan).

Hillebrandt, P.M. and Cannon, J. (1990), *The Modern Construction Firm* (London: Macmillan).

Hillebrandt, P.M. and Meikle, J.L. (1985), 'Resource Planning for Construction', *Construction Management and Economics*, vol. 3, no. 3, 249–63.

Hillebrandt, P.M., Cannon, J. and Lansley, P. (1995), *The Construction Company In and Out of Recession* (London: Macmillan).

Holmans, A.E. (1971), 'A Forecast of Effective Demand for Housing in Great Britain in the 1970s', *Social Trends*, no. 1 (London: HMSO).

Holmans, A.E. (1995), *Housing Demand and Need in England 1991–2011* (York: Joseph Rowntree Foundation).

Holmans, A.E. and Simpson, M. (1999), *Low demand: Separating Fact from Fiction* (Coventry: Chartered Institute of Housing for Joseph Rowntree Foundation).

Holmans, A.E., Morrison, N. and Whitehead, C. (1998), *How Many Homes Will We Need?* (London: Shelter).

Holti, R., Nicolini, D. and Smalley, M. (1999a), *Building Down Barriers: Prime Contracting: Handbook of Supply Chain Management*, Sections I and II (London: The Tavistock Institute).

Holti, R., Nicolini, D. and Smalley, M. (1999b), *Building Down Barriers: Interim Evaluation Report, The Concept Phase* (London: The Tavistock Institute).

Inland Revenue (1998), *Inland Revenue Statistics, 1998* (London: The Stationery Office).

Ive, G.J. and Gruneberg, S.L. (2000), *The Economics of the Modern Construction Sector* (Basingstoke: Macmillan).

Johnston, T.L., Jackson, L.W., Scott, A. and Welham, P.J. (1972), *The Demand for Private Housing in Scotland: A Report for the Scottish Housing Advisory Committee* (Edinburgh: HMSO).

Joint Working Party on Demand and Output Forecasts of the EDCs for Building and Civil Engineering (1971), *Construction Industry Prospects to 1979* (London: NEDO).

Kafandaris, S. (1980), 'The Building Industry in the Context of Development', *Habitat International*, vol. 5, nos 3–4, 289–322.

Keynes, J.M. (1921), 'Introduction to Cambridge Economic Handbooks', in D.H. Robertson (ed.), *Money* (London and Cambridge: Cambridge Economics Handbooks).

Keynes, J.M. (1936), *The General Theory of Employment, Interest and Money* (London: Macmillan).

Knight, F.H. (1921), *Risk, Uncertainty and Profit*, London School of Economics, reprint no. 16 (Boston: Houghton Mifflin).

Latham, M. (1994), *Constructing the Team* (London: HMSO).

Little, I.M.D. (1957), *A Critique of Welfare Economics* (Oxford: Clarendon Press).

Low, S.P. (1991), 'World Markets in Construction: 1. A Regional Analysis', *Construction Management and Economics*, vol. 9, no. 1, 63–71.

Marsh, P.D.V.(1973), *Contract Negotiation* (Aldershot: Gower Press).

Mayo, S.K. (1981), 'Theory and Estimation in the Economics of Housing Demand', *Journal of Urban Economics*, vol. 10, July, 95–116.

Meen, G. (1994), 'Housing and the Economy: Policy and Performance in the Eighties and Nineties', Centre for Housing Research and Urban Studies Occasional Paper 5, University of Glasgow.

Meikle, J.L. and Hillebrandt, P.M. (1989), 'The Potential for Construction Resource Planning', *Habitat International*, vol. 12, no. 4, 63–70.

Mole, T. (1991), 'Building Maintenance Policy: Nebulous, Nevertheless Necessary', in P. Venmore-Rowland, P. Brandon and T. Mole (eds), *Investment Procurement and Performance in Construction* (London: Spon).

Mugume, S. (1999), 'The BOT Procurement Method in Uganda: Successes, Failures and Challenges', paper presented at the 2nd Meeting of CIB Task Group 29 (TG29), 'Construction in Developing Countries – Contractor Development', Kampala, 25–6 June.

Muth, R.F. (1960), 'The Demand for Non-Farm Housing', in A.C. Harberger (ed.), *The Demand for Durable Goods* (Chicago: University of Chicago Press), reprinted in J.M Quigley, (ed.) (1997), *The Economics of Housing*, Vol. 1 (Cheltenham, UK and Lyme, US: Edward Elgar).

NEDC (National Economic Development Council) (1964), *The Construction Industry* (London: HMSO).

NJCC (National Joint Consultative Council) (1994), *Codes of Procedure for Tendering* (various) (London: NJCC).

Ofori, G. (1990), *The Construction Industry: Aspects of its Economics and Management* (Singapore: Singapore University Press).

Ofori, G. (1993), *Managing Construction Industry Development* (Singapore: Singapore University Press).

Ofori, G. (1994), 'Establishing Construction Economics as an Academic Discipline', *Construction Management and Economics*, vol. 12, no. 4, July, 295–306.

Ofori, G. and Debrah, Y. (1998), 'Flexible Management of Workers', *Construction Management and Economics*, vol. 16, July, 397–408.

Ofori, G. and Rashid, A.A. (1996), 'Developing world-beating contractors through procurement policies', in R. Taylor (ed.), *North meets South: developing ideas*, Symposium in Durban South Africa (Durban: University of Natal).

ONS (Office of National Statistics) (1998a), *Family Spending: A Report of the 1997–98 Family Expenditure Survey* (London: The Stationery Office).

ONS (1998b) *United Kingdom National Accounts: The Blue Book, 1998 Edition* (London: The Stationery Office).

Park, W.R. (1979), *Construction Bidding for Profit* (New York: Wiley).

Park, W.R. and Chapin Jr., W.B. (1992), *Construction Bidding: Strategic Pricing for Profit*, 2nd edn (New York: Wiley).

Parry Lewis, J. (1965), *Building Cycles and Britain's Growth* (London: Macmillan).

Porter, M. (1990), *The Competitive Advantage of Nations* (New York: Free Press).

Prest, A.R. and Turvey, R. (1965), 'Cost Benefit Analysis: A Survey', *Economic Journal*, vol. 75, no. 4, December.

Quigley, J.M. (1979), 'What Have We Learnt about Urban Housing Markets?', in P. Mieszkowski and M. Straszheim (eds), *Current Issues in Urban Economics* (Baltimore: Johns Hopkins Press).

Raftery, J. (1991), *Principles of Building Economics* (Oxford: Blackwell).

Ramsay, W. (1989), 'Business Objectives and Strategy', in P.M. Hillebrandt and J. Cannon (eds), *The Management of Construction Firms: Aspects of Theory* (London: Macmillan).

Reid, M.G. (1962), *Housing and Income* (Chicago: University of Chicago Press).

Robbins, L. (1935), *An Essay on the Nature and Significance of Economic Science* (London: Macmillan).

Runeson, G. and Skitmore, M. (1999), 'Tendering Theory Revisited', *Construction Management and Economics*, vol. 17, no. 3, May, 285–96.

Russian Construction Research Group (1993), *Construction in the Russian Federation* (London: CFR Ltd and DLC).

Samuelson, P.A. and Nordhaus, W.D. (1995), *Economics*, International Edition, 15th edn (New York: McGraw-Hill).

Segal Quince Wicksteed Ltd (1995), Latvian Construction and Building Materials Sector Study, Unpublished report, commissioned by the Latvian Development Agency, funded by EU PHARE programme.

Shackle, G.L.S. (1952), *Expectations in Economics* (Cambridge: Cambridge University Press).

Shackle, G.L.S. (1955), *Uncertainty in Economics* (Cambridge: Cambridge University Press).

Shash, A.A. (1993), 'Factors Considered in Tendering Decisions by Top UK Contractors', *Construction Management and Economics*, vol. 11, no. 2, March, 111–18.

Shubik, M. (1955), *Strategy and Market Situation* (New York: Wiley).

Simon, H. (1959), 'Theories of Decision Making in Economics and Behavioral Science', *American Economic Review*, vol. 49, 253–83.

Tam, C.M. and Leung, A.W.T. (1999), 'Risk Management of BOT projects in South East Asian Countries', in S. Ogunlana (ed.), *Profitable Partnership in Construction Procurement*, Proceedings of the CIB W92 and TG 23 Joint Symposium, Chiang Mai, Thailand, January (London: Spon).

UN (United Nations) (1968), *International Standard Industrial Classification of all Economic Activities*, Statistical Paper Series m, no. 4 (New York: UN).

US Department of Commerce (1994), 'Environmental Accounts', *Survey of Current Business*, vol. 74, no. 4, April.

Vipond, J. and Walker, J.B. (1972), 'The Determinants of Housing Expenditure and Owner Occupation', *Bulletin of the Oxford University Institute of Economics and Statistics*, vol. 34, no. 2, May.

Wall, M. (1993), 'Building Maintenance in the Context of Developing Countries', *Construction Management and Economics*, vol. 11, no. 3, July, 186–93.

Whitehead, C.M.E. (1974), *The UK Housing Market: An Econometric Model* (Farnborough, Hants: Saxon House).

Williamson, O.E. (1967), 'Hierarchical Control and Optimum Firm Size', in D. Needham (ed.), *Readings in the Economics of Industrial Organization* (New York: Holt, Rinehart & Winston).

Williamson, O.E. (1975), *Markets and Hierarchies, Analysis and Anti-Trust Implications* (New York: Fife Free Press).

Willig, R.D. (1987), 'Contestable Markets', in J. Eatwell, M. Murray and P. Newman (eds), *The New Palgrave: A Dictionary of Economics*, Vol. 1 (London: Macmillan).

Index

Note: definitions and discussions of the use of terms are in **bold** type.

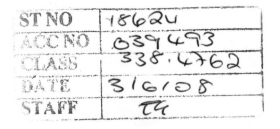